FRIED EGGS AND FISH SCALES

JON TAYLOR

FRIED EGGS

Tales from a
Sointula Troller

& FISH

SCALES

HARBOUR
PUBLISHING

Harbour Publishing Co. Ltd.

P.O. Box 219, Madeira Park, BC, V0N 2H0

www.harbourpublishing.com

Edited by Emma Biron
Text design by Carleton Wilson
Illustrations by Hiromi Ford
Printed and bound in Canada

Supported by the Province of British Columbia

Harbour Publishing acknowledges the support of the Canada Council for the Arts, the Government of Canada and the Province of British Columbia through the BC Arts Council.

LIBRARY AND ARCHIVES CANADA CATALOGUING IN PUBLICATION

Title: Fried eggs and fish scales : tales from a Sointula troller / Jon Taylor.
Names: Taylor, Jon David, author.
Identifiers: Canadiana (print) 20230572723 | Canadiana (ebook) 20230572774 | ISBN 9781990776656 (softcover) | ISBN 9781990776663 (EPUB)
Subjects: LCSH: Taylor, Jon David. | LCSH: Fishers—British Columbia—Sointula—Biography. | LCSH: Sointula (B.C.)—Biography. | LCSH: Sointula (B.C.)—Social life and customs. | LCSH: Taylor, Jon David—Family. | LCGFT: Autobiographies.
Classification: LCC SH20.T39 A3 2024 | DDC 639.2092—dc23

Here's to life, no matter how crazy it makes us.

CONTENTS

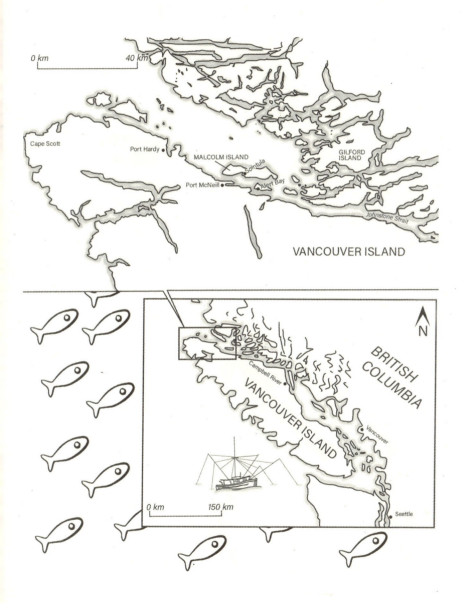

0 km 40 km

Cape Scott

Port Hardy

MALCOLM ISLAND

Sointula

Port McNeill

Alert Bay

GILFORD ISLAND

Johnstone Strait

VANCOUVER ISLAND

N

BRITISH COLUMBIA

Campbell River

VANCOUVER ISLAND

Vancouver

0 km 150 km

Seattle

THERE IS AN ISLAND

Wedged between the mainland of British Columbia and the northern end of Vancouver Island there are myriad rocks, shoals and islands. Toward the southeast the islands cover most of the geography and the arms of the sea run like deep narrow rivers between them, creating their own wind and weather patterns. Toward the northeastern end of Queen Charlotte Strait, just where the groundswell ceases to be a major factor, there is an island.

Unlike its neighbours, it is built of sand and gravel, an artifact of the last great ice age. There is only one other like it in the area; the rest are solid stone. Geologically speaking, it is an unstable structure. Just a pile of sand and gravel dumped where the glacier stopped to rest. Any amount of serious wave action would wash it away and a good shake would let the whole thing subside back onto the bottom of the sea. Even as it is, the wind and tide constantly chew at the edges and erase land every few years like an artist rubbing out a charcoal sketch.

Naturally, the island has no good harbours, no good surface water and not much in the way of soil. To the aboriginal people of the area, it was an inconvenient place to land, but useful for some raw materials. No one wanted to live there for any length of time because it was plain that it was busy sinking back into the sea. About a hundred years ago, some poor Scandinavians arrived on the scene. Since no one else wanted the place, it was given to them on the condition that they bring in more people and make money. They haven't done much of either, but they are still there and nowadays some five hundred of us call it home—though the government calls it Malcolm Island.

My own family arrived on the scene in 1916. They were planning to move there and live a good life amongst fellow expatriates who shared a language and a dream. They had been living on Chichigoff Island in Alaska, where things were rough but reasonable. When they saw what this place looked like, they didn't even get off the boat. They booked passage to Cuba. None of us came back until 1976. In the interim, either the place had gotten better or I had developed standards lower than those of my grandparents. I've been here ever since.

I claim that most people here die of terminal mildew. Even if that's not true, it can be a dark, dank and dreary place at times. It is unlike the other Gulf Islands that the BC chamber of commerce loves to tout. This one is cold and wet in the winter, cool and wet in the summer, and has regular windstorms that would rate headlines if they hit Vancouver. If you manage to find a bit of soil dry enough for a garden, and add enough lime to neutralize the cedar water, you can grow all the cabbage, currants and potatoes you want. If you want corn, tomatoes and beans—move to Courtenay.

For all its obvious faults the Island (as I call it) does have a few redeeming features. It seems to inspire stubbornness and give rise to abnormal levels of individuality. For anyone susceptible to such things it's an unbeatable combination. There is also another power that comes into play at times. For many of us, our first glimpse of the Island came from the deck of the ferry. There is something magical about the view as you come into the landing, and more than a few of us have decided that this place would be home even before we had set foot on shore.

It could be the setting. A neat, cozy village in a protected bay, with a backdrop of snow-covered peaks. Or it could be that the town faces south and only gets half the rain of its nearest north-facing neighbour. But most of us believe that it is simple magic. If the Island wants you, it gets you. What happens after that is your problem.

EMPLOYMENT INSURANCE

For many years the chief occupations on the Island were fishing, logging and drinking. They alternated with the season and the price of fish and timber. Naturally some individuals were more involved in one aspect or another, but virtually everyone did all of them to some degree.

Back in those days the town tended to have its own problems, and its own solutions. Outsiders seldom came around, and if they did they couldn't usually speak the language.

Times change whether we like it or not. Some years ago a wave of hippies swept across the Island, followed by developers and yuppies. They were all looking for paradise, and once they thought they'd found it, they all tried to change it. Most of them stayed a short while causing great amounts of angst and consternation, and then left. Those few who remained were themselves changed as much as they brought change.

Our beloved federal government was directly responsible for a good number of the changes that affected the Island. One of the big changes was employment insurance.

In the old days fishing was a family affair. Typically a man and woman would get interested in one another, and sooner or later she would agree to go fishing with him. If things in the fishboat went well, then nature took its usual course and all went as expected.

Some families fished with babies on the boat, others left the young ones with the grandparents or friends. At one point there was a child-care co-op. Various young mothers pooled their resources so as to care for their own children and a large handful of others. The parents who were out fishing paid to have the children eat at the local restaurant twice a day. The extra kids were divvied up between the various homes and a good time was had by all.

Then the government introduced employment insurance. It was nice to get money for staying home in the winter time, and if you weren't into going logging you could just stay home and collect on your fishing claim. This of course changed the ratio of logging, drinking and fishing in many lives, but it was still the same basic operation.

The problem, and real change, was that you were not allowed to fish with your wife. At least, not if she was to get EI. It seemed silly not to take the money, so something had to be done. The obvious answer was wife-trading.

The Island had a long history of such doings, so it didn't take long to work out the details. Boats that usually fished next to each other were careful to be seen leaving the harbour carrying each other's wives. A few hundred yards outside the breakwater the two boats would pull up beside each other. A week later the two boats would meet just short of the breakwater, then land with the proper crew on board.

Now obviously this wasn't too much of a hassle, but it was nonetheless a lie. And then there were the stories of friends who failed to meet at the appointed time and place, and worse yet the wife's insurance claim was based on the crew share of the other boat. Whether it was higher or lower than the husband's boat, it caused either envy or recriminations.

The next step was obvious. The wives stopped fishing at all. Since they were lying to the government anyhow, why not take full advantage of the situation and leave the little woman at home to mind the children. It got to the point where women who had never seen a boat were listed as crew, and some boats that had only one bunk were listed as having three deck hands.

The government finally saw the light and let husbands and wives fish together, but it was too late. The seeds had been sown. Fishing was good money, but unemployment insurance was the gravy, and more people looked for "pogey" cheques than looked for fish.

A few years ago, back when hippies were the problem, a bunch of us were lined up at the credit union to cash our unemployment cheques—about six hippies with one local old-timer near the middle of the line. The Old-Timer was irritated and, speaking to no one in particular, vented his anger for all to hear. He was furious that he had to wait in line. He was furious that he had to wait in line behind hippies. And he was furious that each and every hippy had an EI cheque in his hand. The fact that he too had an EI cheque in hand apparently had no relevance; besides—he'd earned his!

A few moments of silence followed, whereupon the Old-Timer proceeded to tell all and sundry just exactly what he was going to do with his EI. The Canadian government had just seen fit to institute a one million dollar lottery. The tickets were too expensive for most of us to buy, but not for the Old-Timer. He was cashing both his own and his wife's cheque.

One cheque would go to buy lottery tickets and the other to buy booze. He was going to have a party, invite all his friends, divvy up all the tickets amongst those present, and drink until the numbers came up on the TV. He was sure no one else had thought of the idea so it was guaranteed they would all make money. And if they didn't—well, it would just be proof that the whole thing was fixed so the "Frogs" in Quebec would get it all. It sounded like a good plan to the rest of us, but of course we weren't invited.

The Old-Timer was obviously pleased with himself for having put the rest of us in our place. He cashed his cheques and went off to get his party supplies. I and some of the other long-haired trash went to the grocery store to buy food.

YOU GET
USED TO IT

When I started seining I was thirty-two years old, married and the father of two children. The skiff man was eighteen, and figured that my situation in life was the result of me not knowing any better. He told me he understood completely, and that any time I wanted he could get me a date with a really hot woman. He even told me what to do with her. He said there's a secret that most guys don't know, and I probably wouldn't believe it. But the trick to making a woman happy was to go really slow! No shit, go really slow, take five—even ten—minutes.

I'd been around boats all my life, mostly sailboats, with a few months on a couple of freighters thrown in. So when I moved to a small fishing town in 1976, I figured that fishboats wouldn't be that different. Besides, however they were, I'd get used to it. I scribbled a sign that said "wants to crew" plus a phone number, and stuck it up on the bulletin board down at the harbour. Sure enough, I got a call the next day from a fellow who ran something called a seine boat. He asked me if I could row—yes. Could I cook? Yes. Did I get seasick? No. I was hired. I took my sleeping bag and rubber gear on board the *Zeballos*. She was old and falling apart but had lots of heavy equipment for handling the seine net and the skipper told me the pay was real good.

The skipper had been having a run of bad luck. First off, the company didn't give him the boat he wanted, and then he couldn't get a good crew. Every time we'd pass a deadhead, he'd point and say "There goes f—ing Doug" or "There goes f—ing Dick. Get it? Deadheads!" It took me a while to figure out that these were the names of ex-crew members. There seemed to be quite a few of them. Anyhow, I got used to it.

My job was "beach man and cook." Beach man meant that it was my job to ride in the skiff to the beach, then run to the nearest tree with a heavy rope and tie off the end of the net so it would peel off the big drum and make a trap for the salmon that we hoped were there. Good plan—if only there were fish. I was also the cook. That meant that I had to juggle the timing of meals and sometimes yell back to the boat that someone should take the stew off the hot part of the stove. The system kind of worked. It could have worked better, but I thought we were making money so I stuck at it.

There was a strange mentality about working and drinking in those days. If your body was on the boat, you got paid. You might spend the whole trip in your bunk and not even know where you were. But in just getting there you were considered to have done your part. On the other hand, if you missed the boat, you were out. One boat that we fished beside frequently had a crew member whom I never saw. He was a Lysol and Aqua Velva man—meaning he drank it—but he never missed the boat, so he was part of the crew.

Every time we came in from fishing, the skiff man would ask the skipper if he had a couple of bottles that he could borrow. The skipper always did. It was always a 26 of vodka and a 26 of Canadian Club. The vodka always went down fast, like in twenty or thirty minutes. The Canadian Club was to sip. The next morning when he showed up for net work he would be stumbling and puking, but he would do his job and still have enough Canadian Club to straighten himself up. We usually had the next day off, then the following morning we were on our way again. If we were late on the mornings we pulled out, the skipper wouldn't wait for us. If he said that he was leaving at 08:00, he left. If we were running down the dock, he would just pull out and not look back. For this reason, the skiff man would always arrange to be on board early. Sometimes he was in his bunk, but just as often he was lying in the scuppers or sprawled across the galley floor, wherever his friends had dumped him.

At 02:00 one Sunday morning the skipper got a phone call. The skiff man was in jail in Port Hardy, but could be in Port McNeill by 09:00 if the skipper would pay for the taxi. We went to Port McNeill, paid the thirty buck taxi-fare, and I loaned the skiff man another twenty dollars so he could go buy himself a cure from the local bootlegger. He wandered off, only to return a while later to ask if I had really given him the twenty dollars—he couldn't find it. So I gave him another twenty, and he went and found his cure.

While we were waiting for the skiff man to show up, the skipper had walked uptown without saying where he was going. When he got back from wherever it was that he had gone, the skiff man was in the bunk and the skipper was smiling. He didn't stop smiling for the next three days, nor did he have any words of warning for the skiff man. It made us all really nervous, but we never did find out why.

The skiff man's problems had started the night before in the pub. He didn't like the bouncers there, so he had taunted them by running across the tables, kicking beer glasses and screaming, "LOOKING GOOD!" (this was the trademark of some Hollywood personality of the time, and the skiff man had adopted it as his own). When inevitably he was taken into custody by the RCMP, he was horrified to find that he wouldn't be out of jail in time to make the boat. He knew that he had only one chance to salvage his honour. He broke free from the cops, grabbed a shopping cart from the market next door and rode it to freedom down the steep hill to the ferry terminal, shouting his slogan all the way. The cops went back for their car.

The skiff man hid under the loading ramp until the ferry was pulling out, then jumped onboard. The ferry brought him back and handed him over to the cops. He broke free again, and ran down the ramp and dived into the water. The ferry stopped to rescue him, and turned him over to the cops. The cops handcuffed his hands and feet, and then with great care and sensitivity manoeuvred him into the back of the cruiser. The ferry was long gone but the skiff man wasn't finished yet. He managed to kick out the rear window of the cruiser and was busy worming his way to freedom when he finally passed out.

Back in those days, things were a bit different than now. We fished four days a week. Some of the companies still paid the crew at the same price as the skipper. And the running line wasn't in common use. ("Besides, a good skipper doesn't need one.") This meant that both the beach man and skiff man went with the skiff, even on open sets miles from anywhere. When I suggested to the skipper that one man would be plenty in the skiff, he told me if I didn't want my job there were lots of other people who would be happy to have it. So there were always two of us in the skiff.

According to the rules of the day, a tow (how long you have your net in the water) was supposed to last for twenty minutes. It was explained to me that this was to give the fish time to get in the net and that to tow any longer would be unfair to the boats behind us. The fact that we towed anywhere up to an hour and twenty minutes just meant that the skiff man and I had lots of spare time. We buzz-bombed for coho and humpies (that's fancy pinks to you non-fishermen). The skipper said we couldn't cast toward the net because we would lure the fish out, so we cast away from the net to lure them in. We jigged for cod and looked for signs of the Second Coming. The skiff man explained his sexual exploits in lurid detail. But more than anything else, I just looked at the scenery. You can get used to just about anything, but the scenery here goes on forever and never ceases to amaze me.

Late one summer evening we were setting off the west end of Malcolm Island. The water was oily smooth, and the western sky was an incredible mass of colour. Straight down the middle of Goletas Channel the ball of the sun was sitting on the horizon and the clouds were going nuts. God must have had all the angles working on that one, and I was moved to the point of tears by the transcendent beauty.

The skiff man wasn't happy about this last set of the day. We hadn't got any fish that day and we didn't expect any this time. He felt it was a pointless waste of time and effort, et cetera.

Mostly et cetera. In fact he was so irritated that he was having a temper tantrum. Most skiffs carry a broken oar used to knock the beach knot free—when kicking, cursing, begging and whining haven't moved it. The skiff man was venting his irritation by hammering on the aluminum gunwales with the broken oar while screaming continuous obscenities. I was completely absorbed by the sunset and only vaguely aware of the histrionics going on behind me. When I finally turned around, the tantrum had reached its climax. The skiff man seemed hardly to touch the deck. Both feet in the air, flailing wildly with the oar, screaming at the top of his lungs, he gave one final howl of anguish, fell hard into the bottom of the skiff, pulled the sea anchor over his head and shoulders and sat there quietly whimpering. I turned my attention back to the ever-changing spectacle in the western sky. The skipper picked us up sometime after dark. You really can get used to almost anything.

Two weeks later we found the skiff man dead. Some of his friends had dumped him down the hatch into the engine room. The engine was an old straight-six Cat with individual cylinder cans. The skiff man was wrapped, draped and twisted grotesquely between them. We were supposed to head out that morning and I was there early. I woke the engineer, who was asleep in the forepeak, and we went to do what needed doing. Both of us had seen lots of dead people before and we had no doubt this time. He was cold, grey and stiff, and he was twisted into a position that no living human could have endured. We had to get him out of there, so we reluctantly took hold of him and started bending him into a shape that would let us get him free. His head was wedged awkwardly, so rather than touch the clammy flesh, I took hold of his hair and started to twist. "Oh fuck," he croaked.

We carried him to his bunk. He lay there in the forepeak through the whole of the next three days, and part of the fourth. As we were heading in at the end of the opening, he came up on the bridge to ask the skipper if he had a couple of bottles he could borrow. The skipper was glad to oblige.

WALKING HOME

You'd think that on a small island with only a few hundred people a certain amount of bureaucracy could be dispensed with. Wrong. It seems that since there are so few positions of power to occupy, people fight all the harder to occupy them and then exercise what power they have well past all right and reason.

As a result of this attitude, the town planning council has divided itself up into four separate areas of representation. Area one is the town. That is, the place where the store, café, gas station and school are and where all the important (and normal) people live.

Area two is down by the harbour. The hardware store is there and the population contains a certain percentage of slightly eccentric folk. The total size of these two areas combined is about one mile by two hundred yards and there is absolutely nothing to differentiate one from the other— except, of course, attitude.

The third area is a stretch of road that runs along the beach between one and five miles out of town. The population here believe that they represent the true spirit of the whole Island and in doing so, they see themselves as keepers of the faith. Unfortunately, this usually means trying to prevent a neighbour from doing what he wants to do if there is any suspicion that he might make some money or enjoy himself in the process.

Not that many years ago this area was the local hippie-ville. Many of the same people still live here, but they've mostly lost their hippie papers by having jobs and cutting firewood before it's needed. A great number have sold out to the establishment and even have electric lights and running water. Not like the old days.

Area four really is remote: fifteen or twenty minutes by logging road. And if the people at that end of the Island seem a bit strange, it probably has to do with the brain-shaking ride they have to endure every time they go to town.

Not so long ago the best parties were in the area one to five miles from town. The townies would risk their clean cars on the gravel and pothole road (equal parts of each) in search of a good time. If the time was a bit too good and the party-goers inadvertently parked their car in the ditch, it was just an hour's walk home and the highways department would come along in the morning and pull you back onto the road. Due to automotive failures of one form or another, there was often a good bit of foot traffic going up and down the road in the early hours of the morning. There were no streetlights, guardrails or white lines of any kind to guide you or mark the way. But once you got the feel of it, your feet tended to stay in the ruts and you could step right along, despite not being able to see your fingers in front of your face on some moonless nights.

Nonetheless there were certain points along the road that required a bit of extra care. The bridge did have guardrails, but you had to get on to the bridge before the guardrails were of any use. The bridge was situated where it had been easiest to build, without undue reference to the location of the road. Basically, it was offset about fifty feet from what would seem to be the natural path of the road. If you missed the tight jog in the road, you walked straight into the creek. At low tide you could just keep going and barely get your feet wet. On the other side you scrambled up a short bank and found yourself on the road again. If on the other hand you were unfortunate enough to miss the bridge at high tide, the water was twelve feet deep and you were in for a significant swim.

Half a mile farther on there is an even better trap. The road is so close to sea level that the water in the ditches on either side rises and falls with the tide. The road is flat, the ditches are deep, and there is a wicked "S" turn. The number of people who have walked, ridden, or driven into that ditch at this point on the road exceeds the number of people on the Island.

So picture, if you will, an absolutely black—light-sucking— night. The party is over. The car won't go, and domestic tranquillity depends on your getting there before dawn. You start hoofing it down the road. The gravel crunches loudly under your feet. Too loudly—you can't see a thing, and it would be nice to at least be able to hear a bit.

As you step along you become convinced that someone is following you. That's likely enough, so you stop to wait for them. They stop too. When you walk on—they follow. You call out to them—no answer. You get a bit spooked and so you start jogging, trusting your feet will find their way and keep you on the road.

The footsteps behind jog with you. You put on an extra burst of speed—and find yourself waist deep in a ditch full of seawater. The footsteps behind you stop and there are loud crunching sounds and a contented whinny. The horse that has been following you has found some good grass and will no doubt stay there until it runs out.

. You squish, squelch and drip your way toward home. Your Levis weigh a ton and are frigidly plastered to the front of your legs. Your socks are rolled up into lumps the size of golf balls, but you really don't have many options. There's only a mile and a half to go. The road is wide, straight and dry on both sides. You've got your stride back, and you'll be home in twenty minutes. The distance is reeling off behind you, and you have lost all fear of the night. Long smooth strides, feet on autopilot—almost like flying. Somewhere inside you something is singing, and you are one with the night.

As though from a great distance, you observe your body slam full tilt into a huge stationary object. It is warm, covered with hair and making horrible grinding and grunting sounds. You can feel muscles under the skin flex as it gathers itself. You leap back with an involuntary shriek as hot sour breath floods your senses. Deep in your mind something is telling you this can't be happening, but your body doesn't listen.

A loud bovine "moo" and the strident jangle of a cow-bell fills the night as her ladyship galumphs awkwardly into the bush. Answering grunts and moos fill the night around you. There's a whole bloody herd of cattle standing asleep in the road. You yell, stamp your feet and throw stones into the night. Then you are no longer standing on the road and you have to figure out where the road is—and when you find it, which way is home?

COWBOY

When the major North American banks decided that future profits lay in consumer credit, they inadvertently reorganized our society and caused at least one outsider to move to the Island. People who in the past could never have qualified for a bank loan suddenly received a little plastic card in the mail that told them to spend all they desired now and to pay it back bit by bit—later. The unwary simply went bankrupt. The clever organized their lives so as to remain permanently in ever-deepening debt, "processing" their indebtedness with anticipated raises and inflated dollars. The morally negligent saw a different opportunity, and took it.

There was a fellow in Vancouver. He had a good job, drove a nice car, and even had a bit of status from his expertise. He wore a suit to work and kept regular hours. Due to life's odd twists and turns the fellow didn't really want to be as he was—he wanted to be a cowboy. In the late '60s a person living in Vancouver with an urge to be a cowboy would have been less than seven hours drive from cowboy country to the north and less than twelve hours drive from cowboy country to the east. When his little plastic card arrived in the mail—you didn't even have to sign for it—he took it as a sign. A bit of research revealed that if you went to a bank and asked for a cash advance of less than fifty dollars they didn't check with the central bank. They just asked for your signature and handed you the cash. He made a map showing all the local banks, packed his bags and left. His route from bank to bank—forty-five dollars at each one—ended at the ferry terminal in Horseshoe Bay—another sign. Getting out of town seemed like a good thing to do at this point, so he got on the ferry. From Nanaimo to Campbell River he visited every bank he could find, but then he made a fatal error. Instead of turning around and heading north or east, the Cowboy kept going west. It's an understandable mistake; all the old stories and songs tell you that's the right thing to do.

Eventually he saw the Island, paid for a ferry ticket, and halfway across threw the little plastic card over the side. Because of the slapdash way that the cards were handed out, the authorities weren't even sure who they were really looking for. At any rate they never found him. He's still here.

When I arrived I was looking for a farm to buy. I hadn't found one anywhere else in BC, so I took a little detour on my way back to the prairies to have a look at the Island. They don't have cowboys here and they don't have farms, but once the Island gets a hold of you, that kind of stuff doesn't matter any more. When you arrive here—you're home.

BIG SET

We hadn't made much money while I'd been on the boat, but the skipper assured me that we were going to do okay. It was dog salmon time, and he was the best there was at dog salmon; he said so! "It's all in how you go about it," he said. "No point in taking a lot of little sets, you get a feel for the big ones. You just cruise until you get a chance at the big one, then you take it."

We often cruised for days at a time looking for a shot at a big set. We'd run back and forth between Robson Bight and False Head—all day, every day. Once, we went well into the fourth day before we got the net wet. Another time we were off by ourselves at Foster Island and had four good sets in a row. That is, we caught some fish.

When the fourth big bag had to be helped up with the single fall, the skipper said that now it was time to move to a really hot spot. He was afraid that if someone saw us doing well here, they might beat us to where he knew the fish would really be good. In fact a couple of boats were headed our way, so we made like the birds and flocked off. We did fool them. They stopped where we had been, and proceeded to do a series of small but nice sets. I forget why it was that the hot spot wouldn't work for us that day. I think maybe the tide book was wrong again, or maybe the tide just didn't turn that day.

So it came to what would probably be the last of the dog salmon fishing that year in Johnstone Strait. The skipper said we were really going to do it right this time. We went out four days early and tied up in the sweetest little cove you've ever seen. We put a beach line off bow and stern and floated right in the middle of the bay. We got to know the place pretty well. There were cod under the boat and deer on the beach only fifty feet away. The skipper showed me how the current ran and how the set would go. There was a nice sturdy tree just above the water, so I suggested that if we were to back up to the cliff I could step off and tie up with the web right on the beach.

"That's not how it's done," he told me. And so we waited. Bored, but hopeful.

Late in the afternoon on the day before the opening two buddy boats came and tied up with us. Someone mentioned goose hunting across the way in Port Harvey, so in short order we had picked up our lines and headed across the Straits. We dropped anchor at dark and loaded various people and shotguns into the skiff, which then roared off into the night. They came back an hour later, played cards till dawn, then turned in. We arrived back at our tie-up spot just before the opening. There were three boats ahead of us, and the first one was backed up to the little tree that I had liked the look of earlier.

There were fish everywhere. The skipper kept telling me not to point because the other boats would see what I was looking at. What the hell, they were going off like popcorn all around us, and I never pointed unless I could see twenty or thirty sticking their heads up at the same time.

When the starting gun sounded, the first boat idled away from the tree and made a nice quiet sweep of the bay. We went for an open set just above him. The first boat closed up and started brailing. In fact, he didn't get to set again that night. From where we anchored, we could see his deck lights as he brailed till long after dark. We got one fish in our set. Two days later I cooked it for dinner. It was the only one on board.

HOME COOKING

The skipper of a gillnetter that I worked on had an aversion to cooking. As a result he had special racks made for his cupboards so that his wife could make up all his meals in advance, each in its own separate plate, each plate held firmly in the rack. The skipper also had an aversion to doing dishes. We had a special bin to throw all the dirty dishes in, and after each trip his wife would come down to the boat and cart a laundry tub of dishes home to be washed and refilled for the next trip.

The first time she came for the dishes and found that I had done them, she got so upset that she didn't know whether to cry or get angry. She said it was her job to take care of such things and that I was not to do them. Eventually we compromised. I agreed not to do them too well, so they would still need a good washing at home.

When the skipper found out that I could cook and had no objection to doing so, he was in seventh heaven. Suddenly we had more food than we could possibly eat, and many of his wife's delicious meals were going over the side untouched. At first he wanted nothing but the Mexican food that I often prepared, but then he remembered that his true love was home-canned chicken.

For years he and his wife had been canning whole chickens in two-quart jars. And for years he had been packing them

under the bunk as emergency food. He was always careful to use the type of sealer that had a glass top bedded on a thick rubber ring, held in place by a heavy galvanized spring. He was certain that they would last forever. The reason these jars were considered emergency food was that the chicken was all cooked and could be eaten straight from the jar. The reason that he had never eaten one was that you had to open the jar. But now that he had a cook on board, there was nothing to stop him from enjoying his treat. I thought that perhaps he wanted it prepared in some special way. But "No," he said, "just open the jar and put it on some bread."

The chicken was stored under one of the bunks. The bunk was the usual arrangement—a mattress on a plywood sheet with access holes cut in the plywood. The access holes were small, and a good deal of the storage space was reached by lying on the plywood and groping in the dark. The first couple of jars I pulled out looked okay. But at the skipper's suggestion, I reached farther forward. The next few jars showed some significant colour change, and the ones after that were basically black, with large pale pink blotches.

The jars after that got really scary. In some the rubber seal had obviously failed, on others the jars were cracked. Beyond that there were fragments of jars, partially filled with black gooey slime. I was wearing rubber gloves by this time, but I was still totally unprepared for the disgusting mass that my hand encountered at my furthest reach. With great reluctance I withdrew it from under the bunk, and holding my breath, headed for the deck to throw it over the side. The skipper stopped me with a surprisingly harsh tone and demanded to know what I was doing. The thing in my hands was black, slimy and mouldy, the consistency of rotten rubber, and it smelled like a cesspool. Not wanting to release my held breath, I gestured the obvious.

"You can't throw that away, we might need it," he said in an outraged tone. "That's part of my wet suit. If we get the web in the wheel, I'll have to use it to dive."

In fact it was one of his wetsuit boots, but the rest of the description is still accurate. It was a sticky rotten mass with no sign of a hole to stick your foot in. But the skipper said "Never mind." That if he needed it he'd find a way to make it work, and while I was putting it away I might as well put the chicken back under the bunk. Even though it didn't look too good, he was quite sure that it would be fine if we ever really needed it.

Living where and how we do, emergency food is something that most of us keep around in fairly large quantities. The store, if there is one and you can get to it, isn't always open. The power may be out again, and food costs money that may be in short supply. In good times it just makes sense to lay in a few cases of canned food that can be put in a dark corner and eaten if there is no alternative. Various kinds of canned beans or pasta products last for years. The rats and mice can't chew through the steel cans and if the labels fall off—well, that's what we call a potluck. And best of all it can be eaten cold out of the can if there is no way to heat it or there's just no time. One of my most satisfying memories is of hacking the top off a can of pork and beans and scooping it out with my fingers. It was the best pork and beans I'd ever had and I licked my fingers clean (hey, I hadn't eaten for two days and there was nothing else). There's also just something incredibly rich about knowing that if civilization fails I'll have enough food to last until spring, and by then I'll have enough smoked fish and venison to be able to invite the neighbours in. My plan is to survive on black tea, canned beans, fresh potatoes and all the fish and venison that I want. Lots of us feel that way, but it's the old-timers who understand it best.

There was this one old-time logger who was done for, and he knew it. The Logger's floathouse was sinking, he couldn't get a decent claim, his friends wouldn't help him out any more, and he was tired. There was nothing special that was wrong with him. He was just old. Most mornings it took a full pot of coffee just to wake up, and even then he mostly just stared at all the jobs that needed doing and waited for lunchtime to let him ignore it all for a while longer. More often than not, lunch was followed by a stiff shot of OP Rum. After that the day just sort of took care of itself.

The Logger called in the last favour he could collect and had his camp towed into Port Hardy. The official story was that the camp was there to be rebuilt and then sent out to a great new show. It was already arranged, and as close to a sure thing as a gypo logger could get; the bills would be paid when the logs sold. The tug pulled the camp as far as it could go into the back of Hardy Bay, then the Logger spent the next weeks winching it up the beach at the top of each high tide. He never got as far as he might have, but it was close enough. He used some old timbers to build a walkway to dry land, and then he put his real plan into effect.

The Old Logger had it all figured out. He was sure that if he went to the welfare people and declared himself destitute, living in filth and squalor—never mind that he had lived his whole life that way—they would insist that he move into an old folks' home. He knew enough to emphatically say "NO!"—that he would never do such a thing—and that this would make the authorities dig in their heels. In the end he knew they would prevail, and that is exactly what he had in mind; in fact, that's exactly what happened.

The negotiations between the Logger and the authorities took a while to complete, and this provided ample time to clean out the camp. Anything saleable was sold at discount prices and the cash converted into booze. Anything salvageable was traded straight across for booze, which had the effect of saving the long walk to the liquor store.

What with all the sales and trading going on in the back of the bay, a small group of hangers-on developed. They were mostly of the young and long-haired variety and they were more than willing to help drag out the bits and pieces in return for a share of the profits. The sharing out of the profits generally took more time than the sales and trades, so everyone found the work schedule to their liking, and life was good for everyone concerned. Some of the bits and pieces of machinery sold were useful and valuable, and brought in a considerable amount of cash. The sale of one little crimping tool for seven hundred dollars (in cash!) resulted in a celebration somewhat larger than the usual debacle. Small sales and trades brought in beer. Larger sales meant vodka. But a really big sale meant overproof rum, and in this case, many bottles of it.

It was well into the night of the second day of celebrations, and we were deep into a discussion of the virtues of hard drinking and the life style that went with it, when the Old

Logger turned to one of the greenhorns (as he called them), fixed him with a meaningful stare and asked, "Ya ever been really drunk?"

"Yup." The greenhorn figured he had been.

"Ya ever been so drunk ya pissed yourself?"

"Yep," said the greenhorn, "been there."

"Ya ever been so drunk ya shit yourself?"

Now the greenhorn really hated to be outdone by the old codger. It was, after all, something of his manhood that was on the line. But then again, he hated to make such a disgusting claim. The greenhorn paused a moment too long, and the Old Logger knew he had him beat.

"Shit, you ain't never been drunk, and besides, I done worse than that—I got so drunk I killed myself."

He had in fact given it a damn good try, but being as drunk as he was, he screwed it up and lived to tell the tale. It was quite a few years back, when he had a little one-man show up the mainland. He and his wife and kid lived in a floathouse on the side of a big shallow bay. He could tie the boom right next to the house, and at high tide the steamer could deliver supplies right to the front door. He figured that life was good, and that this was about as close to heaven as a mortal man could get.

From up on the hillside one day he could hear the steamer making an unscheduled stop. He got back to the house just in time to see his wife and kid getting on the steamer. She was carrying everything she owned. It seemed that it was not such a good life for her, and she was off to look for a better one. The Logger was dumbfounded, mystified, confused, and he hurt worse than he thought a human could ever hurt. He got out a bottle of overproof rum and drank it—all of it. When he woke up he hurt even worse and couldn't take any more. He crawled blindly out the front door, cinched a few boom chains around himself and rolled off the edge of the float.

Floathouses have no need of garbage collection or sewage systems. The front door or kitchen window serves as an all-purpose disposal and the tide just sort of floats it all away. At least that's how it's supposed to work. In actual fact a large pile of rusty cans, nameless glop and rotting "et cetera" forms right in front of the door, and an unusually low tide on a hot summer day can set off a chemical reaction that would choke a buzzard. When the drunk, hung over, heartbroken, disgustingly dirty Logger regained consciousness, he was draped over the pile of filth and the nearest water was one hundred yards away at the low tide line. He said that all concepts of pain and misery fail utterly to describe the desolation in his soul when he realized he would have to wait at least six hours to drown, and even that would be a slow, wet, cold and torturous process. He undid the chains and crawled back up the dock.

The wheels of bureaucracy turn slowly, but eventually the Old Logger ran out of options and so got a temporary placement in an old age home. He arranged for a young couple to move into his house "to look after things" and to sell whatever they could. The floor sloped and the roof leaked, but the young couple didn't mind. They didn't have to pay rent, there were still a few bits and pieces to sell, and they knew for sure what to do with the money. Actually they were pretty good about it. They drank or smoked the trade goods, but most of the money went up the hill to the old folks' home when they went for their regular visits.

The shipworms that eat lumber and logs for their sustenance grow big in the back of Hardy Bay. It wasn't long before the float started to go down fast. Soon the storage room was down in one corner, with about a foot of water over the floor. Nobody really knew what was in there but everyone felt sort of responsible, so they started moving the stuff up to the high side of the room. There were a vast number of rusty, worn-out boom chains and broken shackles. There were dozens of rusty axe heads waiting for handles and a huge number of unidentifiable "what's-its." The more interesting "what's-its" started making their way up the hill to the old folks' home for identification.

"Oh, that's a busted cross-threaded whango-dangger. Why, I remember the time when Old Thomas took a top maul to that one to get it free—look, you can see the dent right there."

Busted "rezzifrats" and long stories, all fresh in the Old Logger's mind, dim and ancient to the young listeners. Some of the stuff in the storage room was good and useful, but most of it was junk—interesting, but still junk.

There were three big wooden boxes at the bottom of the pile. The lowest one was half submerged but the other two were dry. They were good-looking boxes, well made and tightly sealed. It seemed likely they could hold something of considerable value. We scraped off the accumulated grease, dirt and rat shit and pried them open. We were all mystified. As far as we could tell, the Old Logger had been collecting rubber boots for the last thirty years or so, and sealing them in wooden crates. He didn't have the whole boot, just the red rubber soles—each one worn smooth, heel cut off, and trimmed neatly around the edges.

Lots of comments were made about how too much time in the bush can do that sort of thing to a man, but that none of us would have guessed he was that far gone. Eventually on one of the trips up the hill someone mentioned the flooding and how we were going to dump a lot of stuff, including the red rubber boots.

"What boots would those be?"

The explanation got him rather agitated.

"Do you think I'm nuts—what kind of nonsense is that—who in their right mind would keep old rubber boot soles?"

One of the guys took him a sample a few days later and came back with an explanation. Back in the late forties or early fifties, times had been pretty tough. No logs, no money, no prospects, and winter coming on, and it looked to be a hard one. The bay they were in was full of salmon, so they borrowed a gillnet and started putting up salt fish. As things turned out they got some work and never ate the fish, but they kept it just in case.

"Don't throw that stuff out; nothing wrong with it, never goes bad. Hard times comes again, you'll be glad to have it."

We left the boxes as they were. The rats and mice walked all over it, but as near as we could see, they never tried to eat any of it. We boiled one of them once to see what it would do. We left it all night on the oil stove in one of the Logger's pots. In the morning it was essentially unchanged except for being a bit pasty around the edges. No one wanted to try their luck with eating it.

Shortly after the Old Logger died, we threw all the remaining junk over the side—including the salt fish. It spread out on the bottom like so many red footprints, and stayed there until it was buried by the mud and silt. Even the crabs and starfish wouldn't eat it.

I guess that for eating purposes that salt fish wasn't much good. But it had kept the Logger going through thin and hungry times, just knowing it was there if needed. Sort of like money in the bank, I guess. You really can't afford to take it out because then there wouldn't be anything to fall back on, so it's still there, just in case you really need it.

HOW YOU USE IT

A friend gave me a shotgun for a wedding present. Yes there were reasons, but not the usual ones. At least I think it was from a friend; it came as an anonymous wedding present. A five-shot, twelve-gauge, short-barrelled, pump-action "Defender" shotgun.

Up to that time I had lived many years in many places, some of them quite violent, and yet I'd never felt the need for such a "defensive" weapon. But now that I was living happily in what is probably the safest and most peaceful corner of the solar system, I got a shotgun and immediately started wondering if I needed more. More guns, more ammo, more protection, a more secure defensive perimeter. In short, I became paranoid.

The damn thing played on my mind. It was always in the same place, safely locked away with the shells in another cupboard. But I would find myself calculating the fastest way to get to it, and wondering if I shouldn't start keeping the lock open all the time. Shortly thereafter the shotgun started occupying a more central niche, and small stashes of shells were hidden behind flowerpots and bookshelves, all within easy reach of the critical defensive positions within the house. Pretty soon I had walking sticks, flashlights and butcher knives scattered all over the house—all within easy reach of possible defensive positions. I was set up to repel invaders, survive crossfire, and either stand and fight or escape into the bush—depending on circumstances and the nature of the forces arrayed against me.

Having waited for quite some time for the onslaught to arrive, it finally occurred to me that the fact that I was ready for it now in no way increased the likelihood this extremely unlikely event would ever take place. I concluded that unless I were willing to install meteor protection and live off my lottery winnings, I might just as well return to my previous, unprepared and vulnerable state. I mean, being vulnerable to a non-existent force isn't really such a bad thing—especially if this pleasant state of un-preparedness leads to happiness and the contented enjoyment of life as it comes.

One day an acquaintance suddenly decided that the bad guys had already attacked his house and were holding his wife and child hostage. He attacked his own house, armed with a staggering supply of weapons and ammunition. Unfortunately, when he got there he couldn't distinguish between his family and the bad guys who kept stepping in and out of the shadows. Everyone survived, at least in the physical plane, but the fellow's sanity never did return.

My toys of destruction were quietly put away and eventually sold to a fellow who needed them more than I did. There was just something about the look and feel of that shotgun that made you want to play the Arnold Schwarzenegger game, and something so palpable about its intent that it made you think the game was real.

By similar infusion of attitude and slight of reality, some of the local boys decided to become big-time pirates. Someone had given them an old speedboat. It was a wreck, but had a look about it that made you want to weave at high speed through an obstacle course. Everyone was broke and too lazy to get a job, so they called me up one day, offered me a few beers and asked me to be a technical consultant.

The plan was to put enough power in the boat that they could sneak into the local booming grounds, high-grade a few top quality logs, and get away towing the logs at high speed. They figured that if they only took a few logs at a time, it would be no trick to pull off. The booming ground was only three miles from the secret pirate headquarters and they planned to work only on moonless nights, preferably in a gale. The rest of the plan involved skidding the logs across a public road and secretly milling lumber in a residential neighbourhood.

Even taking into account the gallons of alcohol that fuelled the scheme, no rational explanation can account for the hair-brained absurdity that they became fixed on. I know it was the boat. The damn thing just looked sneaky and made you want to sneak around pulling off stylishly evil pranks. The boat was eventually stripped for action. The protective but heavy layer of fibreglass that kept the water out was removed for the sake of speed. A few planks were replaced and great thought given to caulking the hull.

A few years later the space occupied by the pirate ship was needed for a get-rich scheme involving free-range chickens. The beer and the guitars came out, a little diesel and matches were applied, and a good time was had by all. Nowadays whenever the salmonberry bushes get too thick and we have to clear them back from the house, we find neat little nests of eggs. They are chalky and grey with age, and so old and dry that you can't even throw them as pranks on Halloween.

Sometimes the urge for larcenous activity has a much more concrete source. Like if you see a bunch of valuables just lying there and there's nobody else in sight; well you just take it don't you? And if there's too much to carry you go and get a truck to haul the rest—right?

So perhaps if you are of the straight and narrow variety and you have a social conscience, you might want to check around a bit and investigate as to why there was a vast treasure just lying there. But if you and your buddies are off on a week-long binge and you are in the middle of absolute nowhere, I mean—what the hell.

For many years now I've had an understanding with the town drunks. I'll lend them the money for a bottle, but I won't do so again until I get my money back. The system works. Sometimes I've had to wait two years but I always get the money back. One day three of the town drunks approached me all together. They wanted more money than usual. They needed three thousand dollars. They absolutely had to have the money, and have it now. The initial explanation was that they wanted to charter a seine boat to go up the mainland for a week. When I suggested it would be cheaper to drink at home, they let me in on the real plan.

They had been out in a friend's gillnetter just a few days before, happily poaching deer and crabs, when they found this copper wire. It was a huge thing, as thick as your thumb (or wrist, depending on who was doing the talking). It ran out of the water on one side of a little rocky point, and back in on the other side. They were intrigued, and drunk enough that one of them took an axe to it and cut it through. It was heavy and they rolled up as much as they could pull onto the gillnet drum before cutting it again.

They brought it into Port Hardy and sold it to a scrap dealer who not only paid cash, but also said he would take as much as they could supply. Visions of wealth danced in their inebriated dreams. I thought they were crazy and refused to have anything to do with the scheme. But they got the money from somewhere because the next day they left for the mainland on a boat that could carry a lot more than they were likely to find.

When they got there, however, things had changed. There were military planes flying all over the area, and exercises of some sort going on in the very bays they were interested in. A young man in a zodiac explained to them that they wouldn't be safe amid the military exercises, and that they had best go elsewhere. They did. Back to the bar. There was never a word in the papers or on the radio, but when the drunks eventually got back to look for the wire, it was gone. Even where it had been visible underwater, there was no trace. They thought of hiring a diver to dig in the bottom of the bay, but like most of their schemes this one also drowned in a bottle.

A long while later I saw a note in some magazine that the "world-wide submarine detection link" had gone dead for a while. And a while later, a group of British military commando "survival experts" stumbled out on a beach not far from there, saying they were lost. But in actual fact, none of us has the slightest idea what it was really all about.

The scrap dealer who bought the wire was right beside the highway, and he left the cable in plain sight along his fence. It was there for some years. Eventually the scrap yard was sold, and everything was hauled away. Only the memories remain, and sometimes I'm not really sure about them. Was it all out of a bottle, or did the fate of the world's undersea naval forces hang in the balance for a few hours, with some of our locals' thumbs on the scale?

Even the most peaceful and honest communities are bound to have a small larcenous element. You know...just a little, and not in a bad way. Certainly not in a way that would hurt someone. There is much to be said for a community and a way of life that lets you leave cut-and-split firewood stacked by the side of the road until you need it. People being such as they are, I have heard of a few instances where a person other than the rightful owner has taken some of the wood home. But more often than not a pile will lie unneeded or forgotten until it rots. Not far from my house is a lovely old stack, forty-two years old, never touched that I know of. It is a lovely mass of crumbling punk, lichen, ferns and mushrooms. I remember when it was pulled up off the beach and stacked. It even survived the very short era of the firewood thief.

About twenty years ago a very troubled city-type person moved to the Island. He couldn't believe that the locals were stupid enough to leave cut-and-split firewood right on the edge of the road where anyone could take it. So he figured that the locals were stupid and would never notice. The place being such as it is, most people knew who was taking the wood, but they figured that he must really need it—it happens that way from time to time. And I'm not the only one who has been down on the beach after dark, cutting enough to keep warm until morning.

But the thief soon made his intentions clear when he started selling the wood around town. The rightful owner of the wood went to the fellow and explained to him that his behaviour was unacceptable, but if he would refill the appropriate piles all would be forgotten. He agreed, but not being a local or very careful, he didn't realize that the woodshed he emptied to refill the piles along the road belonged to the same person.

We live in a wet climate, and tire tracks are really easy to read, especially if they run right across the front lawn. More than that, most of the locals are loggers, or at least very experienced at dealing with logs. And most of us would recognize wood that we had cut for ourselves. The rightful owner of the firewood was renowned for his even temper and gentle ways, but I happened by as the firewood owner was having a quiet talk with the thief, and I watched as the thief stayed up all night loading his belongings into his truck. By dawn's early light the thief, along with his wife and kid, was sitting in the ferry lineup, getting out of Dodge.

Just before the first ferry arrived a few of the local boys showed up to take a leisurely rummage through the contents of the truck—just to be sure that nothing belonging to the Island left on that ferry. The thief was the absolute model of contrite passivity. Whether the logger's peavey that came off the truck belonged here or somewhere else I never knew, but the thief didn't say anything about it staying behind.

Since then, the firewood stacks along the road have been safe enough. I've seen one of the local drunks stop and grab the odd piece now and then. But I've also seen him organize a couple of trucks and a few cases of beer and spend the day cutting wood for all concerned. At the end of the day there is always a small pile of wood left at the side of the road. I'm sure he would never admit that it was there for whomever needed it, but it's still there nonetheless.

FISH SLIME AND PERSONAL HYGIENE

Everyone knows that fish are slimy. There is even a kind of fish called a slime eel. If you put a slime eel in a bucket of water, in a short while you have a bucket of gelatinous slime. Slime eels are sometimes in demand for their skins. Properly tanned they look just like alligator or snake and are used to make boots, purses, belts, et cetera. So the next time you see a pair of snakeskin boots, think slime.

The other thing that most people don't seem to realize is that fish slime is not water-soluble. In her wisdom, Mother Nature decided to cover fish with a wonderful see-through gel that makes the bearer hard to catch and harder to hold. But above all else, it keeps them from getting wet. It's waterproof. You really only have to think about it for a few minutes before the wisdom in this becomes obvious. I mean, if ducks' ass weren't watertight, they'd all fill up and sink backwards. If fish weren't waterproof, they'd dissolve before our very eyes. The only way to save them would be to get them to dry land as fast as possible, and that in itself would create a whole new set of problems.

One of the other things that Mother Nature decided was that slime should attach easily to any other surface, and that once attached, it should stick like crazy glue. Modern chemistry has yet to equal this marvel of tenacity. Once you come

in contact with fish slime, it spreads itself over your whole body. It sticks like glue, and it can't be washed off.

Fish scales and fish slime have a natural affinity for one another. So once you're encased in slime, you just naturally attract every loose fish scale in your area. In our case, that area is a fishboat with a seemingly endless supply of loose scales in every nook and cranny (including my own). There is also a strange technical point to be made here; the stickiest slime is on herring, but the most tenacious scales are on humpies.

Anyhow, two weeks and several showers after a successful fishing trip (meaning lots and lots of slime and scales), I'm sitting at a friend's house having a cup of coffee. As I reach for the milk, a fine spray of fish scales flies from my wrist (or from between my fingers) and decorates the shiny tabletop. Even nature's best glue eventually biodegrades, and for the next few days I will shed like a maple in October. In our community the sensitive host will simply run a damp cloth in your wake and say something like, "Good trip, eh?" In other, more civilized climes, it's more likely that your host will immediately have herself checked for leprosy, and neglect to invite you back for another visit.

When we first moved to Malcolm Island, the house we rented didn't have running water, so we hauled drinking water in buckets from town and took showers at a place built for fishermen. It was in the back of a laundromat and it had an endless supply of hot water. The maintenance on the place wasn't all that good, but the owner was nice, so we took to cleaning the place up a bit before we went in. We could do the floor and walls pretty well with the tools on hand, but the shower curtain was beyond help. No matter what we did to it, it looked and felt slimy. We took to calling it the mucus-lined shower, but we used it anyway. As it turned out, that old shower curtain was like Samson's hair—when it finally went to tatters, the shower was closed and never reopened.

When you've been fishing for a few weeks, you can usually clean yourself well enough to prevent an outbreak of bubonic plague, but the idea of a good hot shower starts sounding pretty nice. First thing I did on tying up in Port Hardy was to phone around and find out which places offered shower facilities. We located one nearby, packed up our clean clothing, and walked to the hotel. The shower room was tiny. There was a chair, a light bulb and a single bed. The shower in the bathroom was one of the old cast-cement types and it hadn't seen any paint for a long while, but we didn't care much—at least hot water came out of the shower head.

My wife took her turn first while I fell asleep in the chair. It seemed like she was in there forever, but eventually I got my turn. The feel of clean, dry clothes after a long period without should be ranked right up there with sex and chocolate. I had dumped my fresh clothing on the bed and now savoured the touch of each clean piece as I put it on.

I was just donning the last of my things when a movement on the bed caught my eye. A closer look revealed more movement. A full scrutiny of the bed cover showed it to be a seething mass of small dark bugs. They were all going about their buggy business in an energetic way, and judging by their numbers had been doing so for quite some time. What I had taken for worn spots on the cover turned out to be seasonal migration routes: spring over by the window, autumn in the pillow and winter between the mattresses. I sincerely hoped that my clothing, which I had so carelessly dumped on the bed, had not blocked any major arterial highways or caused any detours.

When we returned the key to the desk, the lady made some happy comment about our feeling better now. She turned white and began to shake when we told her about the extra guests in the room. We were in such a hurry to get to a drug store that we didn't even think to ask for our money back or for the hotel to buy us the two bottles of Nix that we quickly acquired.

So then it was back to the boat. We stripped off on deck and put everything that had been in the room in a garbage bag. We spread out a big garbage bag on the floor and put a bucket of warm water at the centre. Then we stood on the bag and pulled it up around our necks. Inside the bag it was easy to wash and rinse ourselves without flooding the cabin. Once clean we were off to the laundromat. At the laundromat I wished that I could get into the churning suds myself. The itching had started within moments of discovering the bugs—an obvious psychological effect, which I expected to run its course in a matter of one or two hours. Three days later it was still going strong, and it was spreading. Everyone who heard the story felt the sudden need to scratch. Behind the ears and just under the tops of your socks seemed to be the most affected areas. One friend twenty miles away got it via the radio.

Just trying to write this story has made me acutely aware of crawling sensations on my legs and scalp. It will be interesting to see if it can be spread by means of the printed page.

TROLLING IN THE FOG

I trolled all of one summer in the fog. Day after day I kept my face buried in the rubber hood of the radar. My wife and daughters ran the cockpit. In the mornings it was the Jeannettes and the Browning Islands, dodging tugs and other trollers. In the afternoon it was Mary and Barry rocks, waiting for the evening bite. Most nights found us in Marsh Bay with a handful of other trollers. Sometimes we saw them, mostly we didn't—just green dots on the screen, weaving in and out of each other's way.

We were after coho and springs. We were getting about thirty fish a day, with three or four springs a week on top of that. We were smack dab in the middle of the fleet, both in terms of where we worked and how many fish we were getting. There were days when I had a dozen boats inside the quarter-mile ring on the radar, and never saw one with my eyes. We knew we were catching a middling number of fish because every afternoon a local fish cop would come around to get our hails for the day and tell us the fleet averages.

There was one exception to the averages. One of the boats in our area was getting an average of 130 fish a day. The fish cop was an old acquaintance, and we shared a bit of ancient history with each other, but he never gave a hint as to who the high boat might be. I never would have asked him, at least not with words, but I kept hoping that he might glance in the right direction or nod his head as we spoke. He never did.

I made some new friends on the radio that summer. We recognized each other's voices on the radio and eventually the boats when we passed close enough to see them. I didn't actually meet any of them face to face for several years, but we had lots of time to talk on the radio. They too had heard about the one high boat and we spent a lot of time trying to figure out which one it was and what he was doing right. We didn't succeed at either.

Late in the afternoons the fog would sometimes lift for an hour or two. Green dots on the green screen became shapes and colours, and off in the distance Port Hardy would sparkle in the sun for an hour. Then the fog would come again. On the rare occasion when it was clear, I could see one lone troller working the bottom of the Gordon Group.

Trollers are funny that way. They do well somewhere once and for the next twenty years it becomes their spot. Even if they never get another good bite, it's their secret spot and they will work it every chance they get. A troller I crewed on had a tack at Castle Point based on one day's fishing thirty years before. We never did anything special there, but the skipper kept going back because he had once done well there and thought he might again.

I felt the same way about the bottom end of the Gordon Group, but in reverse. It was a terrible place. The first time I lowered my own cannonballs over the side was there. Twenty minutes later they were there to stay, stuck on what I still say is an uncharted reef. The one lonely boat fishing there was welcome to it. And besides, he was clearly fishing too close to the beach, well over the boundary line. You can't compete with people like that.

The price was pretty good that year and the coho were big. We did one short trip late in the year where every single fish was over ten pounds. The fish buyer was ecstatic and the fish cop told us that we were the high-ranking boat on some of those days—except of course for the high-liner—but he didn't really count any more. He had become a mythic being haunting the edges of our consciousness, but in no way intruding on our reality of fog and thirty fish a day.

I had a good friend who, like myself, had chosen that year to go back to trolling after many years away from it. He was out on the west coast of Vancouver Island somewhere, and what little I heard from mutual friends said he was doing okay. A while later I heard he was doing very well and had managed to pay off his new boat in less than one season. That part may be something of an exaggeration. Nowadays no one pays off a boat in one year—that belongs to the mythic past. And besides, my friend was not the sort of fellow who would pay a debt early—with him it was always late or never.

You no doubt think that you have figured out the rest of the story, and you are probably right. When salmon season was finished, my friend and I met to go cod fishing and inevitably we compared notes. There had been no fish on the west coast of Vancouver Island, but the heavy groundswell had ruptured one of his fuel tanks. So he ran to Port Hardy for repairs and then jumped to the closest place where he hoped he could earn enough money to pay his bills—the bottom end of the Gordon Group.

He spent the whole summer there. Every day was bright and clear, and as long as he kept one pole halfway up the beach, he got about 130 coho a day plus three or four springs. He said that the fog over by the mainland was terrible, but that sometimes in the afternoon he could see a fleet working there and he didn't know how they could stand it. The fish cop had told him that the fleet average was only thirty fish a day and almost no springs. Every morning when he looked at the fog bank he expected to see the whole fleet come steaming out of the mist. But they never did. He said trollers were funny that way—they get hooked on a place and stay there no matter what. I tactfully suggested that his fishing area was somewhat illegal, and that fisheries had no doubt been watching. He said that he thought he was over the line, so he checked with the office. When he asked how close he could get to the beach they told him that if he wanted to put wheels on his boat and roll on the rocks that was fine with them.

He made over $64,000 on the coho and a nice bit extra for the springs.

NOISES

We were fishing the Jeannettes when we got word of an opening in Kelsey Bay. Sockeye! We headed down at once, stopping just long enough in Sointula to off-load our coho and to get a bit of food. Dark found us just coming up to Blenkinsop Bay. We had a thirty-knot westerly on our tail and a strong outgoing tide. What a pile of shit that was!

It had been a long day and I was dead tired already, but the bottom of Blenkinsop Bay is paved with polished marble, or something else just as easy to anchor on. Eighty pounds of anchor, twelve fathoms of chain, and twenty fathoms of line out, and we waltzed across the bay as graceful as could be. The one boat that wasn't dragging said he had a one hundred pound anchor and sixty fathoms of chain out. I wanted to have him declared a menace to navigation every time I slid past him.

It was a wretched night. I had to reset the anchor six times, but the worst part was that every time I dozed off, some little noise would come along to wake me up. The bow roller would click as it moved from side to side. The anchor rumbled as it dragged across the bottom. The winds whistled in the rigging. And somewhere in the boat something went "tick, tick" every time we rolled a bit.

It was the "tick, tick" that broke me. I put dishtowels between the dishes. I dumped the silverware in a heap. I

reached into the food cupboards and randomly stirred the contents. Still "tick, tick." By morning I had moved, wedged, padded and blocked everything I could think of or imagine. I even started to wonder if the boat was falling apart and the sound was really the nails working their way out of the planks. Finally, the wake from a boat running up to re-anchor gave us a bit more of a roll than usual and it all became clear. I opened my toolbox and found two sockets rolling back and forth in their plastic tray—"tick, tick." We caught twenty-two sockeye, and went home.

Skippers in particular seem to be subject to attacks of Strange Noise Anxiety Syndrome. While a crew member might say, "What's that funny sound?" a skipper will likely leap up from his seat red-faced and sweating to scream, "WHAT'S THAT SOUND?" Unexpected loud metallic sounds immediately behind the skipper are unusually effect-ive in creating a quick response.

The wife of one of my friends has an itty-bitty temper problem. Her response to minor irritation might be some-thing harmless, like slamming the silverware drawer. Her timing is such that this event is likely to occur just as her skip-per husband is shifting into reverse while making a difficult docking. Despite numerous attempts to convince her that this behaviour is counterproductive, such incidents continue to occur. One day a particularly unsettling example of this Strange Noise Anxiety Syndrome occurred, involving a meat cleaver, a head of cabbage, a large tin pot and a crying child, and this led to a momentary disruption of domestic harmony. When the crew member in question was brought to task about this occurrence—her status as wife was by unspoken mutual agreement momentarily suspended—her response was to suggest that noise was nothing, especially compared to the fact that the skipper had to sleep sometime and she owned a cast iron frying pan.

Sometimes it isn't necessary to have the crew set you up for surprises because you can do it yourself. Picture a perfect day. Calm, warm, fish are coming on board at a steady rate. Then out of nowhere there's a loud metallic crash. The engine rpm falls to zero and alarm bells start attacking your eardrums. Once the alarm is shut off you notice that the can of nice cold beer you had set down for just a moment has vibrated off the shelf, landed on the throttle handle, shut down the engine and is now soaking into your seat.

Sometimes simply knowing that something could happen is enough to set you off. I had been reading my motor's tech manual trying to figure out some routine bit of maintenance when I came across the "how to stop your runaway engine" section. Sticking rags into the air intake (once the air filter is removed; see page 78) sounded okay, but the idea of carefully disconnecting each injector line while leaning over a big diesel doing twenty grand rpm sounded a bit hairy. So there I am, just putting the boat in gear and pushing the throttle gently forward when, as though in the far distance, the rpm instantly goes up to astronomical heights. The noise was increasing exponentially as I threw off the engine room hatch. I was just reaching for a big wad of rags when the airplane that had buzzed me from behind—yes, that's exactly how I felt—continued on his merry way. The fact that the pilot was a friend just saying "hello" in no way mollified the outrage that I felt, but my engine thumped on happily at its normal speed.

Simultaneous but unrelated sounds are a category all to themselves. You reach for the throttle and the crew sneezes. You touch the radar and the radio gives a blast of static. You reach for the coffee and a can falls over in the cupboard. Some days these occurrences make you think of all the interlocking harmonies in the universe and how there are really no unrelated actions. While on other days you think that somewhere, somehow, someone is trying to tell you something very important and you keep missing the point.

In a world powered by diesel engines, perhaps the most significant sound of all is silence. Picture this: I'm sound asleep in my bunk. Just outside the hatch above my head a flag is fluttering and snapping in the wind. It stops. I sit bolt upright in bed and cry, "Oh, god, the motor stopped!" I'm sure adrenalin is good for something, but it sure does play havoc with taking a good nap.

A friend pulled out of Kelsey Bay late one dark and windy night. It was raining (isn't it always?), the tide was going the wrong way (isn't it always?) and the water was full of stumps and broken limbs. The 671 under the floorboards was humming along happily and the skipper was just settling in for a good thrash when the motor stopped. It didn't flutter, fade or die, it just ceased to exist. No hum, no vibration, no nothing. Five very long seconds later it was back, humming away as though nothing had happened. He said that for those moments it was as though the motor had disappeared from the past, present and future. He said he had the feeling that if he had looked into the engine room—play the "Twilight Zone" theme here—it would have been empty. He had that boat another five or six years and the motor always ran perfectly, but you could see that from then on he never really trusted it. You could see that he always had half an ear turned to it, waiting for it to do its trick again.

VIEWFINDER

The great white hotels come every summer. Giant slugs we call them—great shapeless blobs of glossy steel oozing their way down the channels and between the rocks. Some people call them luxury liners, but I have my doubts. They always arrive in the spring along with the black ants and leave in the autumn with the deer flies. They travel in packs. If you can only see one, look again. Just out of sight in a fog bank or just behind an island there's another one lurking, and when you do see it, it will be aiming straight at you.

I'm sure that, all told, I have spent days of my life just trying to cross Blackney Pass in the fog. It's like playing on the freeway. The only place where I get to turn the tables is at the Parsons Island light. There's a rock there that I can get behind but they can't. When a big one comes around the corner and lines up on me, I get a little giggle deep down inside and feel smug as can be.

I can sit right there with that big bow wave coming right at me and *know* that there's a rock between me and him and that he *can't* pass over it. At the last minute they always chicken out and swing into open water. I just laugh—though I do have a recurring nightmare that at the last moment I look at the sounder and find that I've drifted into deep water and they're finally going to get me.

The strange part of liner behaviour is trying to figure out the why of it all. Sometimes they run fast, sometimes slow. Sometimes they idle at night past all the pretty places, then take off like racehorses at dawn. And most of all, why do people pay thousands of dollars just to look at where they could just as easily be?

My fishing partner and I used to fish small, rather quaint-looking boats, right in the heart of liner and whale-watching country. A liner would come around Cracroft Point and suddenly everyone on deck would be staring at us through a viewfinder. The whale-watching boats sometimes used to go in circles around us, to get all the angles I suppose. Sometimes they would even come alongside and talk to us. Local colour, I guess they called it. Hell, I'll bet there've been hundreds of thousands of dollars paid just to look at us and that much again spent on film and videotape.

One of the whale-watching boats pulled up to my partner one day. There were no whales, so I guess we qualified as a secondary point of interest. From a hundred yards away I saw the flashbulbs pop and watched as my partner tried not to get his lines wrapped in the prop. Until then we had never known what they saw in us, but a few weeks later we found out.

The whale-watching boat pulled up to my partner again and handed him an envelope. Inside there were two pictures of my partner, his boat and dog, and a carefully written note saying, "Dear Mr. Canadian fisherman—I like your boat and your dog very much. Can I come fishing with you next summer?" I don't remember if there was a picture of the young lady who wrote the note or not. I doubt it would have mattered. Mother Nature has a way of getting people to overlook some differences and to remain totally oblivious to others. The differences in this case were something like eighteen inches in height, one hundred and eighty pounds in weight, plus the lack of a shared language or culture. Didn't matter. Letters were exchanged and finally one summer day my partner disappeared. I expected that he would be gone for three or four days. Ten days later I heard he was up the mainland somewhere—fishing, I foolishly supposed. A week after that he was back at work just like usual.

My partner isn't the kind of person who likes to hear (much less answer) questions, so I waited and let the story unfold by itself. Seems there was a chaperone of some sort and hours of time spent piecing together conversations from the dictionary one word at a time. And laughter. Maybe sometimes no one was exactly sure why they were laughing, but it seemed to be the thing to do at the time, and it felt good— really good.

Sometime after that the chaperone announced that she had to attend to business elsewhere. As soon as she left, my partner and his new friend began a tour of the mainland channels and bays, and more importantly, an introduction took place. An introduction to a way of life that most people believe doesn't exist anymore. Small floathouses in isolated bays, no electricity, fresh water from the creek in buckets, and a silence so still it feels like it has been there since the beginning of time. My partner says that the critical moment came when they were halfway up Retreat Passage and his new friend looked around the boat and asked where the bathroom was. He handed her the galvanized bucket. She nodded her understanding. Things progressed rapidly from there.

The boats we fish today aren't nearly as picturesque as the ones we had then, but they are a good bit bigger and more comfortable. For my partner it was a matter of necessity. His new wife could fit into the old boat and did so for a couple of years, but now that there's five of them on board...

And as for me I have a new boat too, but I didn't get mine because of a whale-watcher. It was a radio call from a friend at home. "There's someone here that you haven't seen for a long time. She wants to see you—again." I figured she'd come a long way to get there, so I asked her to stay.

We still scorn the giant slugs and whale-watchers; they're not really working boats like ours. But it is fun to wonder what's going on in the minds of some of the people on deck. Is one of them looking through the viewfinder hoping to see a new life? And will any of them be brave enough to find one? If someone wants to take your picture—smile.

UNDER THE TABLE

One of the local loggers decided to go fishing. He leased a boat, went out, and ran into the biggest pile of sockeye that anyone had ever seen. Six weeks later he was rich. So he bought his own boat. The boat he chose needed a bit of upgrading and some structural work. The logger spoke to his accountant who told him that rather than pay CPP, EI, vacation pay, et cetera, it would be better if everyone worked under the table.

To the logger this meant that the workers were not to tell anyone that they were working for him and that likewise they must not declare the income on their income tax. If they did, he would get in trouble. Once a person has something like that stuck in his mind, there's no use in trying to explain reality to him, so you just go along with it.

In keeping with the clandestine nature of the venture, the logger hired a group of social deviants and misfits who probably didn't know taxes from taxis anyhow. I like to think that I was hired because the first-choice misfit was unavailable and I was the only person in town who could do the job (self-delusion is a survival skill). At any rate, the assembled crew was so far off-centre that no one there was sure reality actually existed, much less in what direction it might possibly lie.

This story loses much of its potency because the actual language used by the Aqua Velva Man (named not because he used it, but because he drank it) was too vile to consider repeating in print or even in this space-time continuum. His only topic of conversation was his need for a woman and his sole term of reference for a female human was a vulgar reference to female genitalia. His mantra would start as soon as he was able to talk (10:30 a.m. to 11:00 a.m.) and would continue until quitting time. "Need me a c—, gonna get me a c—, gonna fuck me a c—." It never stopped. It was a chant equal to any devotee's and just as sincere.

The chief caulker was a vigorous anti-Semite, a pro-Arab, anti-British, pro-IRA atheist, and he spoke about one of these topics constantly. He had a distinctive manner of speaking and a deep resonant voice. He sounded like an old-time preacher calling down fire and brimstone.

A third part-time worker was a newly born-again Christian and his effervescent joy made it difficult for him to say anything but "bless his holy name." Ask if it was coffee time and he would say, "Why, yes it is, bless his holy name." Ask if you should cut off the propeller with a hacksaw and he would say, "That would be wonderful, bless his holy name." He at least had other topics that he could talk about, but all were punctuated by innumerable blessings to his holy name. He also started the radio war.

In our area there are two radio stations: CBC and Old-Time Slime, plus public service announcements, such as, "Anyone having lost a child's sweater at the ballpark last Friday about 02:00 pm with red stripes and a green balloon in the left hand pocket can claim it by identifying it at ..."

I like CBC. But we listened to Old-Time Slime. It made me crazy. I developed a plot to run a fake ad in the local paper. "Wanted: uninformed, incompetent, illiterate nincompoop to read news on local radio station." It would seem to be the kind of ad that most of the announcers responded to.

Bad as that was, the radio war made it worse. Suddenly, instead of local inanity, we would have a devout voice praising his holy name and telling us what a good time we were all going to have in heaven, if only we didn't go to hell first. The Christian was sneaking tapes in and turning up the volume so he couldn't hear anyone tell him to turn them off. When the tape ended he would ask us all for commentary and then suggest several times that it would be a good thing if his tapes would totally change our lives and make us full of everlasting joy, just like him.

A fourth participant in this circus was a "sidewalk superintendent" type, unpaid but nonetheless a constant presence. He was a very unhappy person who felt that his family was always taking advantage of him, or stealing from him, or saying nasty things about him, and he needed to talk to all of us about it—all the time.

In self-defence I would pick jobs down under the boat and wear earplugs. It didn't help much. Aftershave-breath wanted to tell me that he needed a woman. The caulker wanted to know what I thought about the Arab/Israeli conflict. The Christian wanted to know if I had seen the light yet. And the sidewalk super kept saying, "What do you think of that? My own family!"

Somehow quitting time arrived at the end of each day. But just as you thought that sanity might be the next stop, you'd see the logger surreptitiously waving for you to "come over here." He had gotten it into his mind that if we couldn't see the boat from where we got paid, it somehow didn't count. Likewise, if there was no proper payday, we couldn't be seen to be working for him. Since he always had a pocket full of cash, we always followed him. We would talk about the weather, fishing prospects, or blowing up logging machines. (It's something loggers take pride in.) Or maybe tell a joke. Somewhere along the line he would press a roll of bills into your hand, adding the admonition, "Remember, it's under the table."

Money earned under the table is wonderful stuff. You don't have to fill out any forms and it goes twice as far as the other kind. The only trouble is that you have to be crazy to do it.

APPLIANCES

Some fishboats have a toilet. Some of these work, and some of them are used. I have also been on boats where the toilet didn't work, but was used anyway. That's a different story. All boats have deck buckets, they are all used and they all work.

I was crewing on a gillnetter that was owned by a potato farmer. This meant that we had lots of potatoes, but they were all field run—covered with thick dirt and of varying quality. I was out on deck washing potatoes in the deck bucket when the skipper came out to see what I was up to. He watched for a few moments and then said, "Yeah, we need another deck bucket; cook those potatoes real good."

My current boat has two deck buckets and an electric toilet. You flip an electrical switch and the toilet makes whirling and sucking noises. Sometimes it even flushes, usually with the help of a small piece of wire kept strategically nearby. Halfway between Numas Island and Malcolm Point one day our electric toilet slowed to a halt and made deep grumbling noises. Several days and $160 later, it was back to its usual cheerful whirring self. Deck buckets make no noise and cost $6.89 on sale. Our motto: "Bucket and chuck it."

One winter I was working in a boat shop. We were putting a new water tank and a water pump in a local tugboat. I had installed the new tank in the lazaret while the foreman

did the plumbing. He was installing a new Water Puppy for the kitchen sink, but he couldn't get it to prime. We fiddled for hours. We filled the new tank, double-checked for obstructions in the line, but all to no avail. Finally he told me to try the pump in the head while he traced the piping again. The head pump worked. I filled the bowl and the foreman crawled about in the engine room. A while later he came up with a funny look on his face and tried the pump again at the kitchen sink. It worked fine until, with a loud sucking sound, it dried the line to the toilet bowl. For some reason the toilet had been set up to flush with fresh water, but there had never been a check-valve in the system. The head was higher than the pump and hooked in downstream, so every time the sink faucet ran it was pulling from both the water tank and the toilet bowl line. Neither of us had ever worked on that boat, but we both had an irresistible urge to sniff the coffee pot and wondered if they drank it strong.

Virtually all BC fishboats have an oil stove. It runs off the same fuel as the motor, keeps the cabin warm and dry, and provides a place to cook—all with varying degrees of success. My first oil stove took twenty minutes at full tilt to congeal an egg to the point it would slide, not ooze, from side to side as the boat rolled. It took forty-five minutes to an hour to make bacon edible, provided you liked your bacon limp. It also had the amazing ability to burn with a perfect blue smokeless flame and produce no heat whatsoever. Another oil stove I worked with got a new top every year or two because the cast iron glowed red hot on the lowest setting, and pale white on high. You could read by the light it gave off on a dark night. A kettle of cold water put on for tea would leap and hop about like a jackhammer run amok.

Oil stoves never work quite right, and we have all evolved our own favourite trick to keep them going. I clean the filter and carburetor. I got so handy on one stove that I could shut off the oil, clean the carb and filter, and turn it on again before the flame went out. A friend of mine always vacuums his stove to make it burn cleanly. He even got a special 12-volt vacuum so that he could clean out the stove while we were out at the fish camp. One day the stove wouldn't light, so he vacuumed out the soot and about an inch of unburned oil. Then he lit the stove, but it still wouldn't burn clean, so he vacuumed it again—without letting the flame go out. The oil-saturated vacuum sucked up the flame and detonated, filling every inch of the cabin with oily soot. My friend walked out of the cabin carrying the flaming halves of the vacuum and dropped them over the side. The mess in the cabin took days to clean up and even years later you could find odd little deposits of oily carbon hidden away in unlikely places.

Just this spring some tourists stopped by our floats to get some bait and ask where the big ones lived. One of the kids started jigging for shiners. He hooked into something heavy and thought he had a big one. When he wondered aloud how a vacuum cleaner had come to pollute the bottom of this pristine wilderness bay, I told him the whole story. I warned him that if he snagged the electric toilet, he would probably lose his jig, but he kept right on jigging. I don't think he believed me.

The other thing the kid might have snagged off the bottom of the bay was a barbecue. For reasons I never understood, each and every skipper brought his own propane-powered barbecue to the camp, and they brought a new one every year. There was one big float near the centre of the camp where we often congregated, and it was here that the barbecues collected. They were all mostly the same. They were the cheapest model money could buy and for the most part turned to rust in a matter of weeks. Each device started out with a twenty-pound tank of propane and a connecting hose that worked. In a few short weeks some of the propane bottles were empty. Some of the connecting hoses were broken and most of the barbecues could no longer spin or go up and down. The few remaining workable models were cherished and fed salvaged parts from the dead ones.

The reason the barbecues wore out so fast was due not only to neglect and salt air, but also to their being in constant use. Most of the fishermen who lived in the camp were status First Nations who had the right to hunt wild game all year round and felt they had a right to any other traditional food as well. There was a lot more venison consumed in that camp than pork or beef, and various types of fish were always on the grill. I would often arrive back at the camp to be greeted by the smell of venison being grilled and loud invitations to join in the meal.

At the end of the season the weather had become too cold and wet to do so much outdoor cooking. One year someone tried to build a tent over the last few remaining barbecues. First it melted and then it blew away in a gale. I never actually saw anyone throw a rusty barbecue off the dock, but I likewise only remember one occasion when I saw a barbecue loaded back onto a boat to be taken home. It was one of those big fancy ones with large crab- and corn-cooker pots built onto the sides. It also had a one-hundred-pound propane tank attached. As I recall, there was talk of taking it home to deep-fry a turkey.

So I don't know if all the other barbecues went over the side or not. The water there is fairly deep and a rusted barbecue doesn't take up much space, but still. I calculate that about two hundred barbecues went missing from that spot. I would never have the nerve to jig for cod anywhere near that place.

OUTHOUSE

Fish camps belong to a way of life that for the most part ended thirty years ago. But I was fortunate enough to live and work out of one for over thirty years. Every spring we gather up boards, nails, spikes, dogs and maybe a stray boom stick or two and head out to our bay. We drag what's left of last year's floats off the mud flats where we parked them in the autumn and string them all together again.

After a few seasons the regulars have their own floats or tie-up spots. If some stranger parks in your spot, you ask him to move. It's like a private club with visitor parking out front. Most years we have seven to ten permanent residents at the dock, storing their fish and gear wherever there's room. The place is a mess, but it's our mess and we love it.

Unlike most fisheries, we keep our fish alive, storing them in the water in pens. We also seine herring for bait and keep them alive in large holding ponds. In order to make a little extra cash we sell live bait to the sporties at five bucks per dozen. We like to say that each herring is worth twelve dollars to us, because that's what we would get for a perfect cod that we would catch on a live herring. However our customers are seldom impressed with this explanation.

One spring three of us were so broke we decided to move to the fish camp early so we wouldn't be tempted to spend

money that we didn't have. We arrived six weeks before the season, set up camp, and seined up some herring. We set up a big sign that said LIVE BAIT and then we waited. Waiting for your ship to come in is never easy. But the other two guys were smokers, and smokers never rest easy when they see their supply going down, especially when there's no way to replace it. They got so upset that they smoked even more. Just before things got critical we collected enough money from some sporties to send someone to Alert Bay for food and smokes. Really though, there wasn't enough money for both, and I had a pretty good idea what would happen.

Salvation came in the form of tax-free roll-your-own kits from the reserve. Because of that, there was enough cash left over that we got spaghetti too. After that we managed well enough and after six weeks I had cleared fifty bucks above expenses—really not that bad.

One of the facts of living in a fish camp is the need for an outhouse. Everybody is always hanging around in everybody else's boat and most of the boats don't even have a head, much less a private one. So most of us thought of an outhouse as a necessity. Ernie was an exception to that rule. Ernie was a plumber by trade when he wasn't fishing, but he figured that a deck bucket was as good as anything else and privacy wasn't an issue. Ernie's great passion in life was fresh white bread. Anyone going near a store knew to bring him a fresh loaf. He would cut thick slices and make huge sandwiches. If Mother Nature happened to call on him at such a time, and she often did, he would put the deck bucket up on the hatch and sit there enjoying his sandwich while talking to the passersby. The rest of us used the outhouse. My business partner and I built the first outhouse from driftwood and scraps. It served well for a few years, until one of the logs under it rotted away and it became unsafe to use. I knew that I had a few sheets of plywood in my shop at home, so I headed up the line to get them. I pre-fabbed the whole thing, even painted it. I loaded it on deck and headed back to the bay.

One of our floats had been built with the ends of the logs sticking out in such a way that the outhouse fit perfectly. We nailed it together, spiked it in place, added guy wires so the wind wouldn't blow it away and took turns christening it. It worked fine. The traditional half moon was cut in the side at just the right height to put light on your reading material and the door had a two-position latch that offered the user a choice between total privacy and better ventilation. The out-house was right next to my tie-up spot, so every morning I got to greet everyone as they were coming and going.

What with the fish pens in the water and all, we discour-aged mink and seals from hanging around the bay. But of course the odd one wandered through from time to time. As long as they stayed away from the pens, there was no prob-lem. But one morning as I was about to take my turn in the outhouse, I spotted a seal looking hard in my direction. He didn't seem to be in a hurry, and I was, so I continued on my way. I was just in the act of achieving my purpose in being there, when much to my consternation there was a mighty upheaval in the water beneath me. It soaked both my nether regions and propelled me out the door.

When I regained my composure I was standing on the dock with my pants around my ankles, dripping water and firmly clutching my masculinity, making sure it was still there.

(I checked several times.) Looking back at the outhouse, I was greeted by a very perplexed looking seal, his head up through the toilet seat and a look of distaste around his mouth. I don't know what kind of fish he thought it was, but I'll bet he never eats "finless brown" again.

I wasn't the only source of outhouse entertainment. The structure was at the end of a long, skinny float, so anyone going to or from it was considered fair game for scatological commentary or just plain insults of the usual kind. Naturally, a person who needed to get there fast was mostly immune to, if not completely unaware of, the things being said at such a time. The usual pattern was for a person to start getting restless. They might stand up or pace, but eventually their glance would start travelling toward the structure at the end of the float, and then they would start walking that way. Some kind of exponential mathematics is needed to describe what happens next. The closer the person gets to the outhouse, the shorter his steps become, but the faster he takes them. You can easily judge the urgency of the situation by noting the shortness of the step and the speed at which they are taken. I'm sure the science branch of DFO could work out a formula to feed into their latest computer model, but I guess we'll leave that to them.

Conversely, when a person leaves the outhouse, most have a curious habit of stepping out the door, facing away from the edifice and stopping. Sometimes this pause is used to adjust clothing or anatomy, but as a general rule the longer the pause, the more successful the visit. The person then starts with long slow steps that become more and more normal as he approaches the place he originated from.

In case you haven't guessed, there really isn't too much to do while sitting around the camp waiting for the tide to turn, or the wind to go down, or whatever. So, small everyday occurrences tend to take on a larger-than-life aspect. One day, while the outhouse was occupied, there came a great deal of noise and banging from inside. It sounded as though someone were trying to rip the toilet seat off the floor. As it turned out later, that is exactly what was happening. Then the door flew open and the occupant threw himself down on the dock. Reaching beseechingly back under the outhouse he proceeded to paddle vigorously, as if to draw some floating object toward himself. The remark was made that it must have been a real good one for him to want it back to do it over again. But the frantic person ignored the commentary. Instead he jumped up and ran to the herring pen for a long-handled brailler, with which he once again began trying to extract some object from beneath the sacred throne. Again with no success. In a final desperate move, he pried up some of the boards from the walkway and from this vantage point managed to recover his desire—a lottery ticket that had fallen out of his shirt pocket.

We all laughed as he put it in the boat to dry. He said that as soon as the ticket fell out of his pocket, he knew it was a winner. The harder it was to recover, the more convinced he became that this was true. The smell of something burning in his boat stopped all conversation short. "Oh no, oh no! Ouch, ow-e-e-e! Damn!" He had laid the ticket flat on the cool side of the oil stove to dry, but it had somehow blown over to the hot side, and was now charred and crisp. But in fact, the number was still clearly visible, and as he stood there, shirt soaked and fingers burnt, he began to smile. He knew he was going to win—and he wasn't the only one who believed it.

Quite a few cans of beer were used to discuss what he could do with his winnings, what good friends we had always been and what kind of party we expected. Late the next afternoon the ticket, now encased in a plastic bag and taped to a board, was transported to Alert Bay. A tour was made of the pubs and then at last the verdict came: "Not a winner." Zip, zilch, nada, not one fucking number. There was stunned silence for a moment. Then he ordered another beer.

"Just shithouse luck," he said, and threw his ticket in the trash.

CALL HOME

Four of us were down in Victoria. We were sitting around the kitchen table doing some crafty thing that involved a lot of insulting comments and a beer or two. We had the oldies station on the radio and would take turns singing along with something absurd from 1968. It was mid-March—cold, wet and windy (forty knots SE)—but the wood fire kept things nice. It was herring season and someone made the comment that "they were supposed to let the punts go today," and hoped that the weather was better up north.

When the news came on, the announcer started talking about a boat trying to get to Port Hardy and not making it. The table went dead silent. Someone carefully turned up the volume, and we all took a firm grip on our feelings. The ten or fifteen seconds that the announcer took to tell the details went on forever. When the boat's name was finally mentioned, each of us looked at the other and said "no, not me" with our eyes. The announcer's voice went on, mentioning a helicopter rescue.

As time started to run normally again, we each double-checked our memories—that none of ours were on that boat. Mentally unpacking the bags we had packed moments before, we let the relief flow, and mumbled hopes that the crew had survival suits on. Images of heavy spray sweeping over a rolling deck while trying to get the lift cable hooked to the D-ring quickly faded from mind, leaving only a faint queasiness. The fire still snapped and crackled, and whatever it was that made it necessary to badger one another reasserted itself at the top of the agenda. I kept wondering what packer Bill's son was on this year, and someone else phoned a friend that night just to see if he could hear anything strange in her voice.

Nowadays everyone has cell phones, auto-tels, or satellite links and you can call your wife or stockbroker in comfort and privacy. But it wasn't always that way and staying in touch while fishing used to be a problem. The only legal way to call home was on the land-line telephone, but of course you had to be in a harbour with a pay phone to do that. Lots of people put an illegal AM radio in their house and hid the antenna in the attic. Careful calls were made at prearranged times. Calls were kept short and no names were used. The authorities were on the lookout for illegal radios and travelled around in special cars looking for them. If they found one in a house it was confiscated and you paid a fine. The government had reserved those radio frequencies for nautical use ONLY. When VHF radios became available they were easier to hide but they didn't have the reach of the old AM sets. After that SSBs became available and the authorities gave up chasing us. A friend had one illegally installed in his car. We sat in my front yard and talked to a fellow in South Africa.

The first time we pulled into Namu, I couldn't believe the amount of noise. Virtually every boat had their AM radio phones turned up full volume on their deck speakers. As we tied up, our skipper did likewise. At least they all had the same channel on, but the distance between the boats made a deep reverb or echo effect. On top of that, groups of fishermen were sitting around giving catcalls and cheering wildly at times. I was too busy to pay much attention.

We weren't supposed to be there. When we'd left home I'd grubbed for five days and left my wife with no money. The skipper had promised that we'd be back in a few days. At the end of the fourth day, just a few miles from home, he had aimed the boat north and said we were going to Namu. I was irritated. He said that he would call his wife on the radio and have her call mine. He didn't. So there I was, two days overdue as far as my family was concerned, and with no way to get in touch.

There was one public telephone in Namu. It was clearly marked by a sign, and a line of fifty or so men waiting to use it. At 02:00 am the line was still there. When you finally got your turn, you walked into a very large closet and picked up the receiver. There was no dial, nor place to put money, just a receiver. A clear friendly voice at the other end said, "What number would you like?" I gave her the number and was immediately connected. I talked to my wife and kids, said unpleasant things about the skipper, and went into some detail as to the type of recreational activity I hoped to enjoy when I got home.

As I was walking back to the boat I suddenly realized that the conversation I was hearing on the collective radios around the harbour was disturbingly similar to the one I had just had with my wife. Some poor shmuck was up in the RADIO TELEPHONE BOOTH, telling his girl what an asshole his skipper was, and how he was going to eat her panties when he got home. The whole harbour was listening on the radio, and the poor schmuck and his girl were using names.

A few years later, there were various channels on the VHF radio for making phone calls. You could call an operator and get connected, but you had to ask for privacy so people listening couldn't hear both sides of the conversation. If you didn't ask, everyone could hear everything. Some people get off on listening to other people's conversations. But in fact, if you want to make a call you have to listen so as to know when to signal the operator. When there are a lot of people trying to get on, it can become a real zoo. Half a dozen people will try to signal the operator at every break in the conversation, and it can sometimes go on like that for hours. I have a radio where I can reach it from the bunk, so I turn on the phone channel and wait for a break. Some nights that means I get to hear a lot of calls.

One night when the traffic was worse than usual a young couple got on the air. She was in a mall in Campbell River with her little brother and sister. He was in a seiner somewhere up in the inlets. They did it all. They did "shy hello." They talked about their clothes. They talked about their dreams. By that point her little brother and sister were hysterical to go to the bathroom, but the talk went on and on and the kids wet their pants. The couple talked about taking off their clothes and their underwear. Then they talked about each piece of anatomy that the other had. In detail. At about this time the guy said, "Hey, do you know we're on the radio? Other people can hear us."

She giggled and said, "Have we been using names?"

"No," he said.

"Tell me what you'd do next," she said. And he did. Then she returned the favour.

When they finally hung up, instead of the usual rush of people trying to get on the line there was dead silence. The silence lasted several minutes (I think we all had to get our heart rates back to normal). Then finally a voice came on and said, "Gee, I wanted to make a call but my radio is too hot to touch." Nobody wanted to follow that act. The line stayed quiet for at least five minutes more.

PRETTY GIRLS

I know there is at least one advantage to getting older; there are more pretty girls. When I was younger, the only pretty girls were the ones of the right size, shape and colour who were within a year or two of my own age. Since then, things have gotten better.

The first time I noticed this phenomenon I was still young. Nineteen years old in fact, but I had been at sea for many weeks. Our ship docked in Hamburg Germany and we all took a few days off. Mind you, we weren't given the time off, we just went ashore after our last shift and didn't come back for a long time. There were more beautiful girls in that city than anywhere else in the world. Even as I admired them, some rational part of my brain knew that a few weeks before I would have thought of these people as fat, skinny, strange or something else uncomplimentary. But on that day, they all glowed from within. To hell with pretty! These women were beautiful.

Hamburg or Amsterdam, Aruba or Curacao, it was the same. At nineteen, I was the second-oldest deck crew; the rest were fifteen- or sixteen-year-olds. And in every port, they fell in love. We tried to keep an eye on the most vulnerable ones so we'd know where to find them and drag them back to the ship. At nineteen years old, I was the mother hen. In Los

Angeles we missed one, and the ship had to go back for him three weeks later.

As we were pulling into Hamburg, the third officer came around with a clipboard. "Jon, do you want a woman in Hamburg?"

"What?"

"Do you want a woman? We know you haven't been here before, so you don't have a regular, but we'll order one if you want." I declined, but when the ship pulled in there were three taxis of girls waiting for us. The crew greeted their respective partners as the old friends and lovers that they were and exchanged gifts and flowers. As I said, I was young then, and after the first night when I tried to teach the thirty-six-year-old waitress at the Neptune Bar to dance, I reverted to type.

Nowadays most girls and women look pretty, regardless of age, shape or colour. It's the smile that counts, and the way they meet your eyes. I'm not saying that I don't appreciate my version of a sexy girl, or think all the thoughts that go with it. Every time I see a woman who is obviously about to become a mother, I hear this little voice in the back of my mind saying "Gee, I wish I'd done that." I guess my point is that the world is a much prettier place than it used to be, and a lot of the credit goes to the young (in mind as well as body) women who populate it.

I was sitting on my fishboat tying up some cod gear, idly watching the seiner across the way loading the net into the hold, getting ready to go north. The person running the show was a lithe, long-haired twenty-eight-year-old beauty, competently running the hydraulics and flaking the web into the hatch. She was yelling orders to "the boys" and periodically had to run off to solve some problem or other. No question about who was in charge on that show. A week or so later, I stopped to watch a friend pull his halibut gear. His crew is two "perfect 10" twenty-year-olds, one very blonde, the other very dark. Short stylish haircuts and at least, when ashore, clothes that make you remember what it is that you thought you had forgotten about girls.

The skipper was unsnapping the gear and handing off the hooks to be re-baited and racked, all as smooth as could be. When a big soaker broke the surface, two gaffs flashed as one, then it was up and over the rail in less time than it takes to tell. The dark one killed it with a few swift slaps from a baseball bat and the blonde had it dressed just about as quickly. The drum kept rolling and the skipper kept handing off hooks. Should you meet these ladies in a bar some night (and they do enjoy such things) by all means feel free to approach them and try your luck (they definitely enjoy that kind of thing). But a word of caution; be nice, be respectful. Some of those halibut weigh a lot more than you do, and you could be really sorry if you made the ladies mad.

You can always tell when you've been in the fish camp too long. All the hills and valleys start looking like especially desirable bits and pieces of female anatomy. The thick layer of forest covering them makes them look fur covered, and you wonder how it is that you've never considered such a delightful idea before. I got a reality check on that one when I told my wife I wanted to buy a flocking kit and try it out on her. At such times, usually after about ten days in camp, we all start deciding that we need a crew. Mostly, we decide that what we need is a beautiful someone who is incredibly talented, hard working, a great cook, perfect company and just crazy enough to fish quillback rock cod in a bikini. Well, why not? (Actually, there are lots of reasons.) While we're at it, why not more than one crew? After another week, the requirements become less stringent; by then, "human female with a smile" about covers it. There's a prominent British yacht designer who shows all his boats crewed by bare-chested nubile maidens. He even talks about making all the surfaces smooth and corners rounded for comfort against bare skin. If he can do it on a catamaran in the Bahamas, why shouldn't we dream about similar delights on a fishboat in British Columbia? (Actually, there are lots of reasons.)

I have to say in self-defence that, of the few female skippers I've known, all but one always had a male crew and the other sometimes took a female crew strictly on principle.

A friend was broken down in Port Hardy and decided to take advantage of the time while waiting for parts to come in. He checked into the Seagate, showered, got a haircut and a new shirt, and then headed for the pub. It was early in the afternoon when he got there and there were only a few other customers. The occupants of one table, two men and a woman, kept staring at him. About the time he was going to order another beer, the two men came over and asked to sit down. The two of them ordered more beer and paid for it. Then they got down to business.

"What kind'a boat you got?"

What with his showering, shave and clean shirt, my friend tried to bluff.

"What makes you think I'm a fisherman?"

"Is it a good boat? You make some money?"

"Yes, it's a good boat.".

"This is our cousin," he said, nodding to the woman still sitting at the other table. "She needs a job."

The lady stood up and walked over to sit with them, but she didn't say anything.

My friend explained that he didn't need a crew, but the men kept right on talking as though it were a done deal. Finally, the woman spoke for herself. She said that she was clean and a good cook, that she could dress any fish there ever was, and that she worked hard. In fact, that was one of her main requirements—that they—that is, she and my friend—had to fish hard. She said the problem was that other skippers would take her out, get drunk and want to fuck, but never do any fishing.

"Don't get me wrong," she said. "I'll fuck, but we've got to fish too."

The temptation was real. She was good looking and seemed honest. It was like a dream come true. Maybe that was the problem. Dreams are dreams, and when they threaten to become reality, they lose the simplicity and perfection that is so much a part of their appeal. Imaginary lovers never let you down. Real ones are never so reliable.

Anyhow, the parts came for the motor, everything was fixed up, and my friend trundled off to the grounds. Alone. He never called the phone number they had given him, though I know that he kept it for a long time.

The guys in the fish camp are all middle-aged or older. Most of us have teenage or older children and at least one ex-wife. It seems to come with the territory. If you fish for a living, your family life falls apart. I don't think there is one man in the group who doesn't dream of the life he could have with the right woman. Someone who would understand the necessities of wind, weather and tide. And who would respect the skill that we use to keep ourselves alive, and to occasionally prosper.

My second wife fished with me for four years before she announced that she didn't want to be married anymore. While she was there, I was the envy of the fleet. I was working side by side with my best friend, partner and lover. That's heaven right here on earth. When she left, I could just barely work for several years. I'm not really back yet.

If our kids were the right age, most of us brought them along in the summer time so they could see how we made our living. My son and daughter didn't appreciate the lifestyle or living conditions, but they did fine and earned a lot more money than their friends.

After a couple of ten-day trips, my son asked if he could have the next trip off so that he could have a bit of summer vacation. I said that would be fine and that he should relax and enjoy himself. Two days later I got a call from him telling me he had been offered a really good job and was it okay if he took it? I always figure that it's best if kids don't work for their parents. Better they should learn from others. So I gave my blessing to the idea. He called me again a few days later to tell me about the job. He was very enthusiastic. It was the easiest job he had ever had and it paid pretty good. He was working in the fish plant. Dressed in rubber gear the whole day, pushing totes of half frozen fish from place to place through several inches of ice water. But the best part was that they only worked twelve hours a day and he could go home at night and have a shower and still have time to visit his friends. On the boat we were working sixteen hours a day. And no shower.

During the time my son was out at the fish camp with me one of the other guys brought along his two teenage daughters. Everyone was of a similar age, but the expected results never occurred. When my son stopped coming out to the camp the girls expressed some disappointment but not much else. Another brother and sister pair arrived on the scene and the three girls bonded while telling the brother to get lost. The boy was relegated to the company of middle-aged men, all of whom had a work ethic that the boy did not share. He was miserable.

The fish camp is remote enough from the rest of the world that you have to make do with whatever is available. This applies equally to hardware, fishing gear, books and companionship. Most of us spent seven to ten days at the camp before going home for a few days. For some of us, home was only a few hours away, but for others it was a ten-hour ride in a fish truck. For others still the camp *was* home and trips to town were a foray into foreign territory followed by a quick retreat back to the camp. We lived in the camp for six or seven months a year and for the girls it became a summer home. Some of them became proficient at fishing while others got jobs at a nearby lodge. As they got older and felt the increasing need of male company they looked around at what was available and made their choices.

The youngest chose very badly, and became the victim of a pedophile who lived nearby. None of us were certain what was going on, though we were suspicious and talked to the various parents about the situation. The parents weren't concerned. Years later I testified against him. He was convicted and went to jail.

The oldest girl made different kinds of choices. She grew up spending summers in the fish camp and then went home in the autumn to go to school. In the camp she had a constant supply of middle-aged friends. At school she found all the boys to be dull and boring by comparison. She decided that middle-aged men were much more interesting and had much more to offer. The summer after she graduated from high school she arrived at the fish camp and moved in with a man thirty years older than herself. They looked happy, acted happy and were both of legal age, so it certainly wasn't my job to worry about it. If anything, I was just a bit jealous.

The young lady was right into housekeeping, so the boat got cleaner and better kept every day. The fellow decided he might as well please her if possible, so he redecorated the whole interior. A lot of the decoration was on a make-do basis, as money was never a big part of these operations. A roll of red plush carpeting became available, and served to line the forepeak of the boat. Red walls, red floor and red ceiling. Indirect lighting followed, as did a sound system. The fellow had always been a bit of a country-western type of guy, but suddenly it was the sounds of Duran Duran and Def Leopard that emanated from the hidden wall speakers. The whole scene was a bit strange, but in many ways very appealing.

My first marriage had fallen apart during this time, and after a year or so of being alone I hooked up with a very nice local lady who had recently split up with her husband. All was going well in that department until the ex-hubby had a heart attack and my girlfriend decided it was time to go back home and help him. I was devastated. It was the first time in my life that I had a broken heart. I was crushed. Despite the broken heart I had to keep working, and that seemed to be the best cure anyhow. But I was lonely and miserable. Watching my buddy play house with his young sweetheart did nothing for my peace of mind.

One night all the boats ran into town to deliver fish. It was the way we usually did things. The whole fleet would come in at once and a fish truck would meet us at the dock. The off-load would take some hours and end about 01:00 am in the morning. We were always in a mood to celebrate—we were paid in cash—and it was our usual practice to have a few beers together before turning in for the night. There was never enough room at the docks for all of us to tie up, so we rafted together. As it happened, I was tied outside the happy couple's boat that night.

There is a funny thing that sometimes happens when wooden boats are tied together. If the boats are tied in such a way that wood touches wood, any noise in one boat may be heard equally in the other boat. When I walked across the deck of my friend's boat to get to my own I noted that the door was closed, and that a soft red light was leaking out from the forepeak. When I crawled into my own cold and empty bunk, I noticed that I could hear soft music. Lying on my back looking upward, a soft red glow was visible, spread across my ceiling. Then I heard the voices. Soft murmurs, quiet giggles, teasing cadences with squeaks and squeals of delight. I could hear every word and sound as though I were in the same room with them. I felt like I was dying. I turned on my own stereo—J.S. Bach—cranked up Brandenburg Concerto Number Two, and cried (or maybe whimpered) myself to sleep.

While it was all happening I thought it looked like fun, but I assured myself that I could never become involved in such an unequal relationship. I told myself that it just wouldn't be right, at least not for me. Now, looking back, I'm not so sure. They had fun while it lasted and parted as friends. My memories are of being lonely and miserable—theirs are of fun and good times.

One of the other girls expressed an interest in me—an offer, you might say. I said no. I'm not sure why; some sense of moral rectitude, I guess. But I think now that I made the wrong choice. As it was, I was lonely and unhappy that year. It could have been filled with fun and warmth. Whatever was to become of my life after that time, the prelude I chose was cold and empty—but it didn't have to be that way.

Ten years later, on a trip to the south, I stopped at a friend's house, where a young woman was in the process of renting a room. She looked at me for a moment and asked if I used to fish cod out of Double Bay. I didn't recognize her until she told me her name. We brought up all the old stories and she asked if I saw her partner from back then. I did see him from time to time. I asked quite carefully how she felt about things from those days. With total enthusiasm she said that they were the best days of her life. Dating young men her own age was a total drag and middle-aged men had everything she wanted.

Daydreams and fantasies of what might have been are somehow inevitable. From where I am now some thirty years later, I am not sure that I understand anything any better, but I am glad for the memories of what could have been.

CLEAN UP YOUR ACT

Forty years ago, the easiest way for a non-fisherman to know whether the fleet was in or out was to go down to the beach and look for garbage. If you found several dozen ripped green garbage bags spewing their contents along the tide line, you knew that the fleet was fishing. I never understood the urge to seal all the garbage in a bag so as to ensure that it would float. But there were lots of things that I never figured out.

In those days some boats had started compacting their garbage and icing it down to keep it from smelling—that is, a big crew member would stomp on it. But others seemingly took pride in a long string of boxes and bags rolling in their wake. When I surreptitiously put a cardboard box outside the galley door to collect the garbage, the skipper became enraged and ranted about having that shit all over his boat. After that, I made it a point to throw it all over the side, especially when other boats were watching, and particularly the DFO. I only know of one time that anyone complained, and that pleased the skipper; he liked to irritate people.

You could always tell when it was time for an oil change because we lost a couple of hundred RPM off the top end. The skipper wanted more power, so he gave the engineer a quarter and told him to open up the governor and put it between the points. The engineer returned shortly and reported that there were already four quarters in the governor. So the skipper said it was time to change the oil.

Next week when we fuelled up we bought oil. The fuel dock had an oil pump-out system, but the skipper said we'd do it on the grounds. We anchored down the Straits, and after dinner when the motor had cooled a bit, the engineer drained about ten gallons of thick black oil into the bilge and replaced it with new. The motor was noticeably smoother and faster. Then, it being close to dark, the skipper turned on the big engine-driven bilge pump and pumped the used oil over the side.

As it turned out the tide held the oil around us all night and by dawn's early light there was a slick at least two miles long leading straight to us. The first few boats to come by asked if we'd had a breakdown or what. The skipper fired up and walked the boat around the anchor a bit, trying to blow the oil away from us. But it didn't do much good. That night when fishing started, people were still complaining about the oil and hoping not to get it in their nets. The skipper whined as loud as any, but we did the same thing again later that season. However this time we were careful to pump the bilge in the middle of the Straits on a fast outgoing tide.

Another thing that was different twenty years ago was quality control. The law says that hydraulic lines cannot run through the hatch on a fishboat. Really a pretty good idea—they're hot, and if one broke or sprung a leak it would get oil all over the fish. But hydraulic lines on deck have a habit of getting cut or stabbed and this too is inconvenient. If there's not too much oil on the fish, you can wash it off with dish soap. If that won't do, you have to dip them in Varsol.

Sometimes things like that just seem to go wrong, all of their own accord. It was the second opening of the year, and we were at Flower Island. We had been there the week before and had taken about a thousand bluebacks. The company said they wanted more and printed our boat's name on a list somewhere as the top coho producer of the week. The skipper liked that, so there we were, back for more. Only this time the company said they wanted them dressed.

We rolled the first set on deck, and since it was pissing rain we left them there while we went back for a second set. The second set came on board, and we settled in for some serious fish cleaning. Bluebacks are full of plankton and turn black inside faster than you can ever think of dressing them. We figured the rain would keep them cooler than the un-iced hatch, so we just threw the cleaned ones on the hatch cover.

It was about this time that a hydraulic line on the power block cracked open and started showering us with a fine spray of hydraulic oil. What with the rain and all, it took us a while to notice what was happening. By then, you couldn't move around on deck except at a crawl. Valves that never should have been open got closed and we started cleaning up. Each fish had to be dipped in a bucket of dish detergent, and it took hours before the decks were safe to walk on. We were high boat again that week. Two thousand pieces, fifteen hundred pounds. And the crew had to pay for two bottles of dish soap.

Another time we had ice, but it didn't help. The company said it would supply the ice and pay some kind of bonus if we carried it, so we agreed. We went to Alert Bay and got enough ice to fill the grub chest, then ran what was left in the chute into the hold. I had worked at Millards for a while, and knew that some seine boats took huge amounts of ice. It was my job to shovel it into the auger. It was a tedious job, but it paid six dollars an hour, and it was a lot easier if you pulled all the safety grills out of the way.

Anyhow, we had ice, and since two of us had worked on trollers, we shovelled it all into the forward crossing, ready for laying the fish out all nice and neat. For one reason or another we never seemed to get any fish on the ice, so it just kind of sat there, all lonely at the front of the hatch. It must have been cool for a few weeks, because we gradually built up quite a block of ice in the front crossing. A crossing is an area in the hold separated by heavy "pen boards" that prevent the fish from sliding around as the boat rolls—I call it a crossing, though there wasn't a single pen board on the boat. As the ice aged it lost the white look of loose ice. First it turned blue, and then it became clear. The solid block was about as much use as a marble statue, but technically we were carrying ice.

Whenever we went to deliver humpies (fancy pinks) we would pull off the hatch covers and put the fire hose on the fish for a while "to clean them up" so the packer would buy them. Mid-August, a few thousand humpies, and four days in the hold with no ice. We hosed the fish for hours trying to get the smell down. We had to take turns hosing and pumping with the big deck-mounted hand pump. That pump would pass anything if you worked hard enough. I once got half a Stanfield out of it. But that day we had to stop hosing because somehow the shaft tunnel was full of fish, and they plugged the pump.

Three thousand humpies in a thick, warm, pink soup. When the packer said the fish were too spoiled to buy, the skipper said it was our fault for taking such bad care of them. I was broke. I needed the money. I swore we had bed ice and offered to show the buyer. He stuck his head down into the stinking hatch and I pointed at the glittering monolith in the forward crossing. He said that since we were a company boat he would give us UI stamps, but that was all.

The next harbour day, the engineer and I decided it was time for the glacier to go. We gave up banging at it with shovels, and instead chopped it into huge pieces with a fire axe. We used the double fall to pull them out, and put them over the side. The chunks of ice were big enough that we were seriously afraid a boat might run into one of them and damage a plank. Some people did a double take when they saw them floating by, but there was no structural damage done.

The aft bulkhead in that boat was unusually far forward so that the shaft tunnel extended into the lazaret, which was accessed by a hatch in the fish hold. It turned out that someone had left this hatch open. More than a few humpies had fallen through into the lazaret, and thus made their way into the shaft tunnel where they had plugged the pump. We closed the hatch and presumed that the problem would solve itself. Sure enough, after a few days most of the fish rotted enough that they came up through the pump and the bilge returned to its normal colour. We presumed we had solved the problem, but we were wrong. A week later we knew for sure that something, somewhere, was seriously diseased.

We opened the hatch into the lazaret and found that six of the humpies had fallen between the floor timbers. They were a sickly pink, and twice their normal size. None of us had ever smelled anything half so disgusting. Since there were three of us (skiff, beach and engineer) we agreed that we would each remove two fish and clean up whatever mess we made. The skiff man was first. Someone found a red handled spatula in the galley, so armed with that and a deck bucket, he gave it a try. It was like trying to turn a half cooked pancake. He got one and a half fish into the bucket; then he started filling the bucket himself. Each time he reached for the fish, he turned back to the bucket, which was now of course even worse.

For my turn I packed a big chew of snoose in my cheek. (It kills your sense of smell.) I held my breath and worked fast. I came up green and shaky, but I still owned my breakfast. The engineer went into the hold calm as could be, threw his fish in the bucket, and came out like nothing had happened. He was going to make a smart remark, but ran to the rail instead. The dead fish, spatula and deck bucket all went over the side together. No one held the rope.

Even after a shower and a healthy shot of OP Rum, the smell seemed to stick in your nose and throat. It felt for all the world as though the insides of your nasal passages were plated with a thin layer of extremely rancid fish oil, and it just would not go away.

We all went over to my house for dinner that night. The wife had laid on fresh bread and a huge fish stew, which she was slowly stirring with a red handled spatula.

JUSTICE

The old motto in our town used to be "no church, no cops, no trouble." Once we got the church and cops, it was fairly obvious where the trouble came from.

I was cornered on the street one day by a devout but very excitable lady, who gave me five minutes of hell, fire and brimstone before apologizing and hurrying off to her next victim. There was a new preacher in town and he was firing up the Christians. Pretty soon half the town hated the other half for being the wrong kind of Christian, and they had to open a second church. Whereupon some people, unsure of their theology, bounced back and forth from one to the other.

The cops created even more problems. As soon as we got a cop all kinds of things were against the law that we never had to worry about before. If your truck doesn't have head-lights, and as a result you're clever enough not to drive at night, how can it be against the law to drive around in the daytime when no one else has lights on anyway? But once the cop was here he kept finding laws that were being broken, or what he thought was a law being broken.

One night the cop was awakened by the sound of a car crash. A few moments later it happened again, and again and again. Arriving at the scene of the accident he found it still in progress. A small Ford Pinto was hurling itself repeatedly into the side of a parked 4×4. A significant amount of glass and chrome was scattered about, and the radiator of the Pinto was sending up an impressive cloud of steam. But nonetheless it kept right on ramming the 4×4. The cop restrained the driver, who would only say that he was "getting even."

The cop pounded long and loud on the door of the house that belonged to the now-bedraggled 4×4. The door was eventually opened by a sleepy and irate logger. The cop explained what had been happening and said that the Pinto driver would only say he was "getting even." The logger replied that if the driver of the car was busy "getting even," then perhaps the rest of us could go back to bed and just let him get on with it.

In the end justice prevailed. The Pinto driver had to sweep up the stray chrome and broken glass, which he threw in the back of the 4×4. The 4×4 itself then became a trade-in on a new truck, and everyone lived happily ever after.

Intentional car wrecks are not common on the Island, though they are probably more common than accidental wrecks, of which there are very few. Nonetheless, there have been times when it felt like the Island intervened to prevent these, when by all reason they should have occurred.

One time I was riding home on the ferry. Just as we approached the dock I saw a big Cadillac lumbering down the road toward the centre of town. The Cadillac was loaded with a vast number of shake blocks, causing the back end to drag on the ground and throw sparks. Just before it went out of sight behind the library I saw the hood fly open and bend back over the roof. Then the Cadillac disappeared behind the library, where I knew there was a sharp and nasty blind curve. I expected the Cadillac to end up either sideways in the road or firmly embedded in the library wall, but much to my surprise it reappeared, hood still bent over the roof, and proceeded through town to the area where they were delivering shake blocks.

I had a friend riding in the car that day. When I asked for the details he told me that the hood went up with a crash, but the driver simply slouched low in his seat so as to see out under the hood and said "No worries, I got her." He never even slowed down.

And then there is the local legend. The details differ with the teller but everyone swears the essentials are true. There was a long-time resident of our town who couldn't seem to stop his truck fast enough to prevent the front end from looking like an accordion. He tried hard to stop hitting things, but all he did was get frustrated. Sometimes he got so frustrated that he would slam into reverse and do similar damage to the rear of his truck. I don't actually recall him ever owning a pair of headlights or a tailgate.

The fellow's frustration level typically ran pretty high, so when he got a new truck he took some safety precautions. He extended his front and rear bumper supports about two feet beyond the sheet metal, as well as mounting massive iron pipes in place of the usual soft fragile chrome. He then proceeded to brag all over town that absolutely nothing would hurt his truck this time.

Come Saturday night at the town dance, he was being a bit obnoxious about the supposed invulnerability of his truck. Then, in a fit of chemically induced stupidity, he decided to comment on the looks of the bouncer's wife. Or more precisely, "Hey, your wife has buck teeth." The bouncer grunted in a noncommittal manner, smiled patiently and waited until the obnoxious fellow wandered away. The bouncer then excused himself from the dance and walked the few blocks to his house to get his backhoe. He drove the backhoe to the dance, hammered the new truck flat and then returned to his duties as bouncer. It is said that in actual fact the front and rear bumpers didn't have a scratch on them. When later challenged about his actions it is widely reported that his only reply was, "Send me the bill."

TENDER

When I was a kid I wanted my own boat. I wanted it kid-sized and I wanted it to be mine. We already had a "big boat"—all twenty feet of it—that we kept on a mooring in the harbour at Santa B. There was a water taxi that would take us out to it and pick us up when we wanted to go ashore, but I wanted my own skiff and imagined I'd use it to go back and forth whenever I wanted. Happily for me my father felt the same way, so we got out the Sears catalogue and ordered a do-it-yourself boat-building kit. In those olden days you could order anything from a catalogue. Farm tractors, engine rebuild kits, whole house kits, virtually anything. We ordered a kit for an eight-foot pram: a small boat, flat on both ends and suitable for beaching or leaving tied up at the dinghy dock when not in use. Skiffs, prams, dinghies—there are subtle differences, but they are all smallish and often stored on the decks or davits of larger boats. We set up some sawhorses in the garage and put the pram together as instructed. It turned out fine and only leaked a little.

Once we had a pram, whole new worlds of exploration opened up. We could take the big boat to someplace new and then use the pram to explore the places where the big boat couldn't go. People always seem to want a bigger boat, but the bigger the boat the fewer places it can explore. Depending on one's resources this cycle can eventually involve several boats and several skiffs. For us poor fishermen in relatively small boats, one skiff is usually enough—but it has to be the right one.

A local crab fisherman figured out that there were a lot of crabs in very shallow water—so shallow that he would need a skiff to run his traps. He asked me to build him the skiff. I agreed, but when he found out that it would cost him five hundred dollars for me to do so, he asked me instead if I would help him build himself a skiff. And oh, could he borrow some tools and a set of plans? In the end the skiff floated and rowed well. The rowing was important, because the Crabber couldn't afford an outboard. So he tied the skiff behind his boat, and he went crabbing.

The next time I saw him there were big patches of duct tape all over the side of his big boat. I naturally asked what had happened. "That skiff you built me is no good," he said. "It's too fast and too long."

In response to my queries, he explained that every time he rowed back to his boat he was going faster than he should have been, and the skiff was so long that the bow always rammed into the side of his big boat before he had a chance to turn. I nodded sagely and let it be.

A short while later the skiff showed up at my shop in the back of a pickup truck. The Crabber asked me to build a small waterproof bulkhead about eighteen inches back from the bow. I did so. Then he borrowed a chainsaw and cut off the front seventeen inches. He covered the rough cuts with roofing tar, and nailed a piece of used plywood over the mess. Then he went crabbing again.

I saw him in town a few weeks later. I asked him, with much smirking and disingenuous interest, how the new skiff was working out. In dead earnest, he replied that it was a much superior boat now. It travelled at the right speed, and it was short enough that he hadn't rammed his own boat once.

Despite the improved skiff, the Crabber decided that crabs were a pain and that the future lay in separating American tourists from their American dollars. To this end he made a deal with a local yacht owner. The yacht owner would be responsible for getting the customers from place to place; the Crabber would bring along a skiff, and would be responsible for getting the paying customers to and from the beach, and for guiding them between the bears and mosquitoes.

The places they were planning to go made towing a skiff something of a poor bet, and due to space limitations the crab skiff couldn't be made to fit on deck. The Crabber therefore needed a new skiff, and so approached me for a suitable design. I had fun designing the new skiff. It was safe, able in rough water and pretty to look at. The Crabber looked over the lines and then decided that what he really needed was a six-foot punt. The lines were a close approximation of a mortar-mixing tub, and it took me three hours and fifteen minutes to build it.

The Crabber was a stout fellow and he tested the punt by rowing himself and his Saint Bernard out into a nasty southeast gale. The dog panicked and went over the side, but the punt never shipped a drop of water, so it was declared seaworthy, safe and able. The first commercial test of the punt involved three adults and a large picnic lunch (read, twenty-four beers), rowing in through the surf at Cape Caution.

A week later the punt was back in my shop for improvements. The Crabber said it didn't float high enough in the water when it had a load on. So he asked me to fill in all the unused spaces with foam blocks. I somewhat naturally assumed that this was for more support in the event of a swamping—but no, it was intended to make it float higher in the water, and to lift faster to approaching waves. When I sarcastically suggested that I might be able to put enough foam in to make the punt hover just above the water, he explained that any foam added below the waterline would make it float higher, but any foam above the waterline would just weigh it down—that there had to be water on the outside of the boat even with the foam, in order to make it act as flotation material.

He took the punt back to Cape Caution, and reported afterwards that the foam made a real difference. He said, "When a big wave rolled up from behind, that extra foam made the punt lift so fast, it damn near flipped ass-over-teakettle." He fixed that by having me remove some of the foam and seating the heaviest passenger at that end of the punt.

The last I heard of the punt, it was for sale in the paper, advertised as an "unsinkable, un-capsizable, great sea boat—six feet four inches long." He sold it for twice what I had charged him to build it. The blunt-nosed crab skiff was on its way to becoming a catamaran. But instead it became a raised bed in his wife's vegetable garden. The last time I saw it, the thimbleberries were in full bloom.

Another time, I was working for a local fisherman, doing repairs and maintenance on his boat in preparation for the upcoming season. There wasn't enough room at the public docks to spread all our bits and pieces around, so we moved the boat over to a private net-float, where we could make all the mess we needed and not worry about cleaning it up for weeks to come.

The net-float was in a quiet corner of the harbour, but there was no walkway from it to the shore. The day we moved the boat over we asked a friend to shuttle us the fifty feet between the net-float and the regular dock. It was obvious that this system of going back and forth wasn't going to be appropriate, so I threw a small skiff in the back of my truck and took it down to the harbour.

The way the floats were laid out, the net-float was only about fifteen feet from the far end of the parking lot. To move around via the usual walkways involved several hundred yards of walking, and then a short boat ride. I parked at the far end of the parking lot, carried the skiff down the rocks, and simply pushed off in the right direction. I didn't even have to row. One good push and the skiff would glide over to the float, and I could then grab the rail and tie up. I kept the oars in the boat in case I needed them, but I never had occasion to use them.

There was a lot of work to do, but Jimmy was busy elsewhere so he seldom came to see what was happening. On the few occasions that he did come by, he would walk out on the regular docks and look across the fifty or so feet of water at what I was doing. I offered repeatedly to bring him over in the skiff, but he was always too busy to take the time. He just kept telling me to carry on, and that I was doing a good job.

One morning I arrived in time to see Jimmy being ferried across to the net-float on his friend's tugboat. The two of them looked at the work I had been doing, and then the tugboat owner said he had to leave. By this time I had brought the skiff across and was standing with them. Jimmy made his friend promise to come back for him later—it seemed a strange request at the time, but I didn't pay much attention.

We worked for some hours and then, as always happens, we needed something from the shop. Jimmy's shop was only half a mile away, but he was obviously concerned about getting there. I had my skiff and my truck right there, so I couldn't see what the problem might be.

He walked over to the skiff and looked at it. "How are we going to get to the beach in that?" he asked. The skiff was very small, just a punt, obviously a second cousin to a horse trough. Nonetheless I had carried heavier loads than the two of us, and we only had about two-and-a-half boat lengths to travel. The skiff was arranged with a fore and aft straddle seat which would allow the load to be distributed in such a way that the skiff stayed relatively level. I got in and moved well forward to allow lots of room in the back for my passenger. Jimmy stood back ten or fifteen feet and kept muttering, "It will never work." I waited a bit while some internal debate raged, and then he acted.

Jimmy was a big man—two-hundred-and-thirty pounds—and extremely strong. I also remember that on this day he was wearing very heavy steel-toed boots. He sprinted for the skiff and, running full tilt, leapt into it. He took two quick steps on top of the seat, flung his great steel-toed boot past my head as he leapt over me, planted one foot on the tiny foredeck, and gave a mighty lunge for the beach. Had he moved just a bit slower he would have made it dry. The force with which he hit the boat propelled us toward the beach at great speed. In the few moments that it took him to negotiate the length of the skiff and make his mighty lunge, we had travelled more than halfway to the safety of the shore. As it was, he landed about thigh-deep on one leg and about knee-deep on the other. From there it took only a single step for him to arrive on dry land.

His final lunge had of course pushed me and the skiff all the way back to the net-float, and as he looked back at us he was furious, and started screaming about how unsafe all skiffs were and how I didn't know how to operate this one in particular. I, on the other hand, was more than a bit miffed over the mistreatment of my skiff, not to mention the fact that his great steel-toed boot had missed my forehead by a less than satisfactory margin.

As we pulled into his yard, he pointed out a large rowboat that sat perpetually by the driveway. It was a sturdy vessel, fourteen feet or so long, wide in the beam and flat bottomed—as stable as a rock, many people would say. Jimmy said that it had nearly killed him, so he now refused to own or operate any skiff of any sort. The rowboat by the driveway was in no danger of going to sea in the near future. Like other skiffs before, it had been filled with dirt and used by his wife as a flower planter for some years.

The trouble with skiffs had started many years before. At that time, Jimmy was a young man just getting started in fishing. After fishing a series of leased boats, he finally decided to have a new one built. In those days it was possible to go down to the shipyard, make a deal, and get your new boat in a few months' time. There just wasn't that much to do; wood was readily available, the craftsmen were ready and able, and the only machinery required was the motor—usually an East-hope. It all went together quickly and painlessly.

Jimmy took delivery of his new boat well before fishing season, so he decided to do a bit of timber prospecting. The boat had been launched from the yard in a light condition, and the yard had then ballasted the boat with beach stone. It was explained to Jimmy that the stone ballast had to stay in until the wood "soaked up" the seawater and started acting as natural ballast.

There was something about the look of all those stones in the hatch that irritated Jim. The boat didn't go as fast as he wanted, and he was sure that hauling around all that weight was costing him fuel. Nonetheless he followed the yard's advice and left the stones in place.

The yard had told him that the stones had to stay there until the wood soaked up, but they hadn't said how long that might be. Fishing season was getting closer, and on his last prospecting trip up the mainland Jimmy decided it was time to see if the boat was ready or not. He was miles up a protected inlet and the water was absolutely smooth. He was afraid to throw the stones overboard—in case he found that he needed them—but he had to find out if he still needed them. He came up with a plan.

Jimmy leaned over the side and put a pencil mark at the water line. He rigged a line to the wheel to keep the boat going in a straight line, and then set about getting the stones out of the hatch. He was still afraid to throw them over the side—in case he needed them—so he stacked them on deck.

By the time he had finished the job and had all the ballast stacked on deck, he was many miles down the inlet and nearing the open water. He checked his pencil mark and found that the boat had not risen even an inch. He was pleased that he could at last get rid of all those obnoxious stones. He had just set about throwing it all overboard when he encountered the first hint of groundswell. Loaded as it was, the boat rolled heavily to one side. All the ballast shifted to the low side where it was prevented from going over the side by the raised bulwarks. The boat was pinned, and it began to take on water into the hold. The boat was about to sink.

With water pouring into the boat, Jimmy wrestled his skiff off the back of the boat where it had been sitting for several months, and pushed it into the water. In the excitement he failed to hold on to the tie-up line. The motor was still running, but with the boat in imminent danger of capsizing and sinking, Jimmy couldn't think of any reason to go below and turn it off. His skiff was rapidly getting farther and farther away, so he took a dive off the back of the boat and swam to the skiff. He just barely made it. When he got to the skiff he went to the wide transom and pulled himself in. As he did so, he was met by a wall of water. The wooden skiff had dried out completely as it lay on the back of the boat, and had sunk within moments of being launched. Jimmy dragged himself in and tried to stand up. The swamped skiff sank beneath his weight, and proceeded to roll about under his feet like a burling log.

In the meantime, the boat was leaning more and more onto its side and started going in circles around Jimmy as he clung to his swamped skiff. Finally the boat was tilted so far over that the ballast stacked on deck slid over the side, and the boat partially righted itself. Jimmy could hear the engine accelerate, as the carburetor was once again level. The wheel was still tied, so the boat ran straight and was going right past him. When the boat had been well over on its side the water level had not touched the motor, but now that it was level the flywheel picked up the water coming under the bulkhead from the hatch, and proceeded to short out the ignition—right in front of Jimmy. Jimmy paddled his sunken skiff to the boat and got on board. He said he nearly gave himself a heart attack bailing out the water with a five gallon bucket. He didn't finish bailing before the boat had drifted into the area of fairly heavy groundswell and was now rolling its rails under. He was sure that the excessive rolling was caused by the loss of the ballast, so he left enough water in the boat to keep it down to his pencil mark. Nonetheless the boat was terribly unstable. He dried out the ignition, then gingerly turned around and made his way to the nearest logging camp. The camp was several hours away, but he made it.

At the logging camp it took him three days to finish drying out the boat and collect more stone ballast. By the time he got home he was a total nervous wreck, but he had figured out what the problem was—it was the skiff. If he hadn't tried to use the unreliable skiff, everything would have been okay. From that day onward, for the rest of his life, he refused to carry or use a skiff.

Jimmy told the story with great sincerity. I don't believe he ever understood what had happened that day.

HAYWIRE TIGHTWAD

I got a job on a troller once, long before I owned my own. I'd never been on one before, so I didn't know what to expect. Trolling, as it turned out, was fun. You lowered down some lines and pulled them up with a fish on each hook. Seemed like a good system to me. Periodically the skipper would come aft, smile, and leave me alone. Once, he even brought me a sandwich. It seemed like a good way to make a living, so I decided to become a troller.

I had never been on a troller before, but I had been on boats all my life and also spent some time as a mechanic. I was able to tell that this particular boat was somewhat less than first class. The gurdies, which wind the lines up and down, were belt-driven off the motor. The drive shaft was a very rusty piece of rebar mounted in one-inch water pipe "bearings." The clutch was a rope—really a series of half-rotten ropes and knots—that you pulled to tighten the drive belt, and the belt-tightener was a clothesline pulley with a notch cut in the frame so you could slip it on the belt. The motor was a venerable Gray. When you moved the throttle up or down you had to go below and reset the timing manually. Likewise, if you were going to spend any time at cruising speed you had to adjust the mix-screws in the carburetor. I'd had lots of cars in the past that could have benefited from such attention, so it was nice to be able to do it while under way, rather than having to pull over to the side of the road.

The skipper was a notorious tightwad—in case you hadn't guessed—but was generous with praise and good stories, so we got along really well. When we grubbed up in Port Hardy for an eight-day trip to the Yankee Bank, I asked for some potato chips as a treat. He bought one bag and said that it would last. It did—just.

Going up Goletas Channel I lowered the trolling wires to tie knots in the weak places. The rule of thumb was that if only one strand was broken, you ignored it. If two strands were broken, you kept an eye on it. If three strands were broken, then you had to cut it and tie a square knot. Once I found two breaks about three feet apart. Rather than repair it twice, I threw out the piece in between. Wrong choice. He never said a word, but the look of shock and mourning on his face was such that I never did it again.

I had been looking for work before the season started and had gone to the skipper's house to suggest hiring me to repair and caulk his skiff for him. About the time we were crossing Nahwitti Bar, he suggested that I might want to caulk the skiff now. Being crew, I wouldn't get any extra pay for doing so. I asked where the cotton and irons were and he handed me a polyester rag and the bread knife from the galley. He told me to tear the rag into thin strips and press them into the seams, being careful not to break the knife.

It was the end of a perfect day when we dropped anchor behind Cape Sutil. The anchor line was rather used looking, but the skipper said it didn't matter—we'd have to jog any-way if the wind came up, so we wouldn't be putting any strain on the anchor. To say that the setting was perfect is a gross understatement. The sea was glassy, the swell minuscule, and the sunset a minor masterpiece. There were a handful of trollers anchored around us. We sat on deck drinking tea and listening to the soft shushing of the gravel beach. After a long silence we threw the dregs of the tea over the side and went below.

As in most older boats the bunks were narrow, but long enough that there was room to keep your things at the head or foot of your bunk. As I started to lay out my sleeping bag, the skipper sternly informed me that I was not to roll around or squirm. Even though my bunk was only a few inches above his head, I thought he was overdoing it. In response to my questioning look he said, "That way the hooks won't get in you."

I guess my non-comprehension showed, so he went on to explain that since he was usually alone, he used the extra bunk to lay out gear. Inevitably some of the hooks got stuck in the mattress, and since he was unwilling to cut the hooks or rip the mattress in order to get them out, they were still there. I used the glow from my cigarette lighter to inspect the mattress. It was a nearly solid mass of rusty fishhooks, many with the spoons still attached.

Before I could suggest turning the mattress over, he said that this was the good side. I rolled out my sleeping bag and slid in with extreme caution. He was right, I did lie still and I never got a single poke. In fact, I worked on the boat for some time before I figured out what was really happening. The skipper had admired my sleeping bag and when I went to take it off the boat, it was fastened to the mattress like Velcro. The skipper thought I might want to leave it where it was, rather than rip it getting the hooks out. I had no qualms about getting the cutters and cutting off the hooks. The skipper just looked kind of sad and said it didn't matter anyhow, the hooks were probably too rusty to use.

Like I said before, the skipper was a tightwad. I never did find a piece of gear on the boat that wasn't second-hand, borrowed or reclaimed from the dump. Every time we passed a kelp island he would study it to see if there was anything good in it that we could salvage. His favourite trick was to snag things on the stabilizer line without getting it tangled in the trolling gear. Then it was my job to get the stabie in, dump the kelp, and save the "good stuff" while we kept right on trolling.

We were in the middle of salvaging a scotchman one day, when I looked at him and said, "I heard a story about you once." He held up his hand and said, "It's probably true." It had been many years ago, when he was young and strong. He was trolling alone five or six miles west of Cape Sutil and there was this incredible kelp island full of scotchmen, lines, glass net-floats and other items of great value too numerous to mention. He went in circles around it for quite a while, but couldn't figure out a way to get at the stuff. Just seeing all the good things there going to waste made him kind of crazy—more so than usual.

In the end, he tied the helm over a few degrees, took a line with him, and dived over the side. The boat was happily going in circles around the kelp island and he figured he could go tie on to the pieces he wanted and climb back aboard when the boat came by again. He said that the kelp was a lot thicker than he had thought it would be and that it was hard to swim from place to place, but he nonetheless got what he was after and got safely back to the boat. He said it was only later that he realized what he had done and he figured that for a while he had been properly out of his mind. It wasn't the only time.

The skipper was outrageously good with a rifle. He had lived his whole life hunting and fishing for food. Any game that he could see was as good as cooked. When I was working for him he was in his seventies, had one gimp arm and double vision. But I never saw him miss a shot. Any floating garbage was worth a .22 shell. I think it was the only extravagance he allowed himself. A blue shark anywhere near the gear always got a hole in its dorsal fin.

He told me that just after World War II, the "Japs" (meaning Canadians of Japanese descent) weren't allowed to fish on the outside of Nahwitti Bar—by order of all the local fishermen. The "Japs" of course figured that such a policy was discriminatory, and to make their point they got a bunch of boats together and came out as a fleet to demand their rights. The skipper said that it looked like it would be war all over again. That was fine with him. To make his point and express his disapproval of their presence, he pulled out his rifle and shot the trolling bells off the first boats he came to.

Unfortunately for the skipper, the politicians had decided that any Canadian could fish anywhere he damn well pleased, and knowing that some fishermen might not feel the same, they sent along the RCMP to see that it happened. The cops boarded his boat, took away his rifle and told him to join them on the police boat. The skipper pointed out that his boat leaked badly and was likely to sink if he wasn't on board to man the pumps. So they took the distributor-cap from the motor and started the long tow back to Port Hardy.

The skipper made the best of the situation by pulling out his accordion and serenading the fleet as he was towed in. He said that there was a popular ballad at the time called "The Prisoner's Song," and he played it as long as there was anyone to hear. By the time they got to Port Hardy, the afternoon wind was doing its usual trick of making Goletas Channel untenable for man or beast. The police boat cut tight to Duval Point, the tow rope broke, and before anything could be done the fishboat was on the rocks and a total loss. The skipper figured that the police owed him a boat, but they told him that it was his own fault that he was being towed in. The judge wouldn't even count the value of the boat toward the fine he got for shooting the bells off the "Jap" troller.

NEW AGE
FEMINIST

The trouble with newcomers to the Island is that they want dry firewood. And they want it now. If they've been around for a while they'll have a few dry pieces in the woodshed, and you can burn just about anything if you have a bit of dry wood to put in around it. But newcomers don't usually have any dry wood, and worse yet, don't really know how to use it if they do.

A friend showed up one day with his chainsaw and peavey in the back of his truck. He announced that we had to go cut some wood for the new schoolteacher—and it had to all be dry. The new teacher was an attractive young lady of the unattached sort, and she had sought out my friend. She had announced that she had no wood and she was cold. There was also some mention of the need for a bigger woodshed. It's like the mating call of the loon—new woman on the Island—needs firewood—wants to be warm and dry.

Getting firewood for yourself can often involve a fair amount of work. After all, you know how much you're going to need all winter and you can't pretend that you didn't know. Cutting firewood for someone else is different, especially if it's a charity case involving a young lady.

The basic idea is to impress the lady with your good will and honourable intent, along with showing off your manly generosity and your skill with a chainsaw. This is best accomplished by locating a dry log just big enough to cover the bed of your pickup truck, at the same time not involving any splitting or heavy lifting. After all, you don't want to show up at her woodshed all sweaty and dirty. Likewise, you don't want to give her too much wood. If you and the young lady hit it off and she is suitably impressed with your efforts, then you need an excuse to invite her out for a day of woodcutting. Too much wood in the shed means no such excuse. On the other side of the coin, if you don't hit it off at least you haven't wasted a whole day cutting her wood.

We worked a bit too hard getting the wood, but my friend was out to make an impression. It was all dry, and it looked like it would keep her going for a good long time. But in fact we both knew it would burn so fast that she would be out of wood in a week or so.

When we arrived at the woodshed, it was about half full, and there were a good number of tire tracks leading to and from it. It was clear that the new teacher didn't quite under-stand the rules and had told more than a few people of her desperate straits. As we pulled in she looked rather sheepish and apologized for having overstated her need. Before we finished unloading there were two more trucks waiting to dump their loads.

The new teacher was a pretty good-looking woman but that wasn't the only reason that everyone was showing up with wood. She had said that she NEEDED wood; so people were bringing it. As we were pulling out of the yard another truck was pulling in. We all had a good laugh about the fire-wood and didn't think much about it until the teacher's car broke down.

She phoned me up one day and asked if I could come over and give her a hand fixing her car. When I arrived, there were already people there working on it, and more were arriving. Hell, there were so many of us that you had to stand in line to get a look under the hood.

When we compared notes it turned out that she had phoned each of us individually to come over and help her fix the car. It was an old Rambler that she had salvaged from the scrapyard. Somewhere, she had read that a good mechanic didn't need to buy new parts or expensive things like that. A real mechanic could just rebuild what was broken, or at worst manufacture what was needed from leftover scraps.

The lady was trying to find out which of us were TRUE mechanics, and she was adamant that she was learning to do the repairs herself. To this end, she insisted on standing in the way and handing out the wrong wrenches. When it was suggested that she go make some coffee, she went up in smoke, ranting about male chauvinism et cetera. But she did stay gone long enough for us to get the car going. We never did get any coffee.

Over the next few weeks I noticed quite a few cars coming and going from her place, but since there were no obvious signs of a party I didn't stop by.

The lady eventually settled for a regular fellow, who hung around for a while, but I guess in the end he just wanted too much and she settled for another. It turned out that the lady had a few rules of engagement about which she was very strict. Aside from various sexual requirements, which she discussed freely, her partner was free to cut the firewood if he wished to—after all, it warmed him also. But she considered herself responsible for half the wood and she would get it in whatever way she saw fit. When she did arrange for firewood she stacked it on her side of the woodshed and made sure that only she brought in pieces from that side.

Likewise, just because her partner was a mechanic didn't mean that he should or could work on her car for her. He was allowed to—if he wanted to—but he was not required, or even needed, to do so. She was, after all, an independent woman and she would, and could, take care of such things in her own ways.

The new schoolteacher lasted for a year or two, but like most transplants to the Island, she got tired of the endless rain and short summers. She tried to fill in her time by starting various women's consciousness groups. But when she ran out of conscious women that were willing to talk to her, she left.

When people give up on a dream they have different ways of moving on. Some just pack up in the night and leave on the first ferry. Others find a new dream and try to convert all their old friends to their own new way. The schoolteacher went on vacation (something schoolteachers are good at) and just didn't come back. She went to Hawaii in June and failed to show up for work in September.

In October she dropped in to pick up a few of her belongings and to sell her half of the firewood. She came by for coffee (I made it) and told me that she had found religion and a new way of life. I'd never seen her look so good, or so calm—and she was laughing about the person she had been in the past. She was quietly and calmly sincere about everything she was doing—and it looked good on her. She left on the first ferry the next morning. We never heard from her again.

JJ

Geographically speaking, the fish camp isn't all that remote. In a straight line it is only about six miles from a fishing community of about 1,500 and only about five miles from a logging road that eventually connects to the rest of the world. But both of these are on the far side of the Island and across a narrow but often rough piece of tide-swept water. From the camp itself, except for the floats you stand on, you can see no manmade object.

The bay where the camp sits faces north, across the Queen Charlotte Straits toward the Coast Mountains, which are snow-capped or snow-covered, depending on the season. Impossibly remote on crystal clear days, or hanging silent and cold just above you on moonlit nights, the mountains create a sense of presence that tugs and pulls at your senses like the pole star draws a compass needle.

The bay itself is a small bay in a larger one, a tiny indent carved by the ice sheets that shaped and polished all this land. Here the glacier lingered on its retreat, filled the back half of the bay with fine silt, then hurriedly departed, leaving the front half of the bay too deep to safely anchor in and the foreshore too smooth to tie to.

The larger bay that holds the smaller one is big enough to have a name, but the small one is not. Despite eighty years of almost constant occupancy, it has an indefinable air of impermanence. Everything here—floathouses, docks, barges—has always been temporary. Most likely it is the winter wind that makes it so. Visually the bay would seem to be protected from the east, but it's not. The hills and islets that surround it form a funnel that throws the full weight of the winter storms on to this one spot. Two hundred yards away the trick works in reverse, and the water may be glassy smooth, even in the biggest blows.

I have seen pictures of this place back in the thirties. Seven floathouses occupy the bay and a community of faces stares out at the camera with a small sense of pride. They built a dam and reservoir back up in the trees and brought water into their cabins. They planned to stay. They're not there now, and the reservoir is a barely discernible depression in the ground. Even though we knew roughly where to find it, it took several years of seasonal wanderings before one of us happened upon it and recognized it for what it had been.

Twenty-five years ago there were two wooden barges moored across the mouth of the bay. Seine boats would tie on the outside to do net work, while trollers and gillnetters tied on the inside, taking turns sleeping in the night or day as their fishery demanded.

The fisheries have died out and changed since then. Twenty years ago, when a few of us needed a place to base our new live cod fishery, we moved into the bay. There was no trace of any former inhabitants.

In the summertime, between salmon openings, there were as many as twenty-one boats fishing out of our small corner of the world. Some boats day-fished locally and returned every afternoon to tie up in their regular places and store their fish in permanent pens. Others fished as groups, and went out to other areas to fish for a few days at a time. But all brought their fish back to the bay to be stored and sold.

The buyer came right to the camp and paid cash on the spot. It was a busy place. So busy at times that we would have to keep one person in camp to buy fish from other fishermen and sell fish to visitors who dropped by. It was difficult to have any time alone—at least in the summertime.

The autumn was different. The tourists disappeared in September and the salmon boats, accustomed to making big money on the salmon, couldn't be bothered hanging around for the little bit extra they would make on cod. The skippers all said they would be back after a hunting trip to the interior, but few—if any—ever came back late in the year. For the cod fishermen it was a different story. It was a bread-and-butter fishery and every day you fished meant a bit more for the larder.

When the weather gradually turned stormy we would sit out the first blow or two tied up in our usual places. But when the next big one was forecast, many of us would head for home to wait it out. Once you were home, it got harder and harder to return to the camp. Even when you wanted to, it wasn't always possible. One October we all went to Alert Bay to wait for better weather. Despite being ready to fish every day, we only fished seven days of the whole month.

Late in November the days are very short, and the weather is usually terrible. Even if you have good weather to fish, there is so little light that it just doesn't seem worth the effort. You'd be tied up reading books by 04:30 every afternoon and you couldn't head out until 08:00 the next morning.

There was one (and only one) fisherman who always stayed until the bitter end of the season. He was different from the rest of us in several ways. All the other fishermen had boats big enough to live in and all the boats had most of the comforts of home. Some weren't very roomy but they all had a cabin to sleep in and a stove to cook on. The exceptional fisherman, JJ by name, had neither. His boat was sixteen feet long, without so much as a scrap of canvas to protect him from the rain and spray. Nothing in the boat ever seemed to work right. The electronics in particular seemed to have a curse on them.

The one thing the boat did have was speed. The motor was old and feeling its age, but it was 150 hp and the sixteen-foot boat could do close to sixty knots. Even with short hours of daylight and sudden changes in the weather JJ could get to where he wanted to be and get back before things got too nasty. In my own boat a trip to Alert Bay took at least an hour, one way. JJ could get there, buy a few supplies and be back in forty minutes.

To make up for the lack of accommodation in his boat, JJ built himself a shelter on an unused log boom that was tied in the back of the bay. He built a framework of 2x4s and stretched blue tarps over it. There was just enough room for a bed—a foamy on a piece of plywood—and a Coleman camp stove. Light came from a Coleman lantern. In the cold damp of winter the heat from the stove and the lantern drew moisture out of the water-logged float, and the cold wind condensed it on the inside of the plastic shelter. I tried to talk him into putting an air vent in the roof to drain off the moist air, but he was afraid of losing what little heat he had and chose to live with the condensation. When the wind came gusting through the bay the shelter would shake and rattle and spray a fine mist of water over everything in sight. In cold weather the condensation froze on the plastic and a gust of wind would spray the interior with a hail of tiny ice pellets.

JJ suffered from serious arthritis, and due to the constant wet conditions in which he worked and slept he had to live on prescription painkillers. He had a house, a wife and two children in Comox. But he said the price of fish was good in the winter—so he stayed.

To kill the long hours of darkness and the gale driven days, JJ took to reading. Most places that cater to fishermen have a pile of paperback books that are either free or so close as to make no difference. You take a pile out, you bring a pile back—it all works out in the end. JJ's reading habits were totally indiscriminate. He would only check that none of his latest choices were recent repeats; then he would read the pile from the top down.

Despite the appalling conditions in which he lived, he seemed content. He had fixed up his domain with a stylish outhouse, an outside storage area and a mooring slip for his boat. For a while he had a radio to listen to. He listened to the Port Hardy station in the daytime, and then at night when the reception got better he would listen to country and western music from Vancouver. In a previous life JJ had been a radio technician in the Canadian Forces, so when for no obvious reason the radio started emitting loud howls and screeches instead of music he put it by his bed with the intention of fixing it—later. Nonetheless he was more at home in his shelter than in his home in Comox.

The day started normally. JJ had an early breakfast and then spent most of the day seeking the elusive rockfish that his living depended upon. Late in the afternoon he dropped his catch in his pen in the bay and then headed to Alert Bay for supplies. Coffee, cigarettes, boat fuel, stove fuel, lantern fuel, canned spaghetti and another pile of books. He arrived back at his float just at dark with the rain and wind coming on strong. Books and groceries went into the shelter—the rest could wait for morning.

JJ fired up the lantern and the stove. The canned spaghetti went into the frying pan and he took a book off the top of the pile. As the wind and rain increased to their usual winter ferocity JJ settled in for a long evening of entertainment.

If at that moment some airborne sentient being were passing high above, it would have seen a mass of small islands in a wind-whipped sea, surrounded by silent white peaks, and at the centre a small but bright tent-like shelter emitting an eerie blue light. Further investigation of this area's sole light source would have revealed a fisherman taking his ease, completely unaware that his life was about to change forever. The book on the top of the pile was by an author that JJ had never heard of—Steven King.

A few hours later, things were not going well at all. It was a classic "dark and stormy night" and the darkness outside the little pool of blue light was full of strange and menacing noises. The float began to creak and groan as the swell picked up and the heavy gusting wind was shaking the whole shelter like a terrier shakes a rat. JJ decided that his only choice was to keep on reading; he needed the distraction of a story to prevent himself from thinking about what he had read so far.

It was shortly after this that the lamp started to run out of fuel. It didn't run out all at once. It dulled down to an orange ball of burning gas, flared to full brightness, then sank to a pale blue glow, before fluttering out. In the darkness the hiss of escaping pressure was muffled by the sound of JJ's heart pounding in his ears.

Being a careful camper, JJ had just that day purchased more lantern fuel. But it was still out in the boat. Out in the storm, the dark and strangely violent storm which was now filled with the presence of an unknown and unseen mind— one that was aware of his tiny blue light, and aware that it had just gone out.

JJ had just enough self-control left to grab his flashlight, intending to look for the fuel. He never made it. In groping for the flashlight he inadvertently turned on the radio, which instantly filled the darkness with the squawks and shrieks of hell. He jumped into his bunk, pulled the sleeping bag over his head, and whimpered until morning.

The next day I was at home still waiting out the storm when I got a call. If I got down to the camp, could I take the rest of his stuff home with me and store it for the winter? And if I got over to Beaver Cove, could I bail out his boat? But it didn't really matter—the motor and electronics were already safe in his garage in Comox.

The last few years that JJ worked late into the season, he was never quite the same. When the days started getting short he would start coming around trying to make sure that we weren't all going to leave at the same time. We would always promise that we were going to work the season right to the bitter end and that we would wait out the storms with him right there in the bay. But it never worked out that way. Sooner or later we would all be gone for a day or two, and when we came back his camp would be empty.

JJ's old skiff is still fishing out of the small bay, but nowadays it sits on the deck of a sixty-foot seine boat most of the time and the man who works it heads in every time the weather report looks glum. JJ took up truck-driving for a while before deciding to go back to his original trade of repairing radios. He stays in Comox year-round now.

INTEGRITY

If there is a common source for community friction it is probably firewood. Well, actually it's chickens, but I refuse to get involved with the small-minded ignorance that dominates this particular arena. I like having a yard full of chickens and anyone who feels otherwise is wrong.

So firewood—the local rule is that wood on the beach, before it is cut or tied up with a line, belongs to whoever gets there first. Those of us who live on the beach tend to feel a bit possessive about logs right in front of the house, but then again, if we really want a particular piece, all we have to do is go tie a line on it, and it's ours. Nonetheless, some people get choked up when someone else cuts wood right in front of their place. I was visiting such a person one day. We were sitting in the kitchen enjoying an afternoon libation of the Scottish persuasion. I was pleasantly relaxed, but the lady of the house was getting more and more upset. Every few minutes she would get up from the table and go look out the living room windows toward the beach.

"They're doing it again," she said in a hiss. I could see that there was a local fellow and his family down on the beach. They had a 4×4 and several chainsaws, and they were working their way along the beach in our direction. It's a good system for getting a truckload of firewood in a hurry; and besides, the next tide will probably fill in any gaps you might create in the log jam that typically covers the beach. My friend and his wife, however, didn't like the system at all. As far as they were concerned it was their wood that was being taken, even though they seldom cut any for themselves, preferring to buy it.

After another round of Scotch, the husband excused himself and headed for the beach. From one hundred yards away we could hear him ranting and generally making an ass of himself. He came back proud and strutting, but I noticed that the woodcutters kept to their previous course. A short while later the truck was jammed to capacity, and they drove off toward the nearest access road. My friend was still simmering, but he felt that he had done his manly duty—and then the truck pulled into the yard. Four young men jumped out of the truck and came up to the house. The spokesman quietly informed my friend that, since he was so adamant that the wood was his, they were delivering it, and where did he want it stacked?

My friend had forgotten that the woodcutters were devout Christians who took their faith very seriously. My friend stuttered and stammered and turned bright red, but the boys threw off better than a cord of wood, and stacked it neatly beside the driveway. Then they went back to the beach and cut another load for their grandmother, whom they had been working for in the first place.

Somehow the sense of manly accomplishment evaporated, and my friend didn't seem to enjoy himself all that much for the rest of the day. For me it was magic. Poetic justice is a rare event in the world and seeing it first hand is even rarer still. Firewood seems to often bring poetic justice to light in this cold wet part of the world. It makes me think of the time Molly built me a fence in the rain while I sat inside babysitting. It's a convoluted story, but it makes some kind of sense in the telling.

It was another North Island pisser. Rain on rain, rain the week before, and more rain to come. I was sitting by the kitchen window drinking hot tea and babysitting the three kids, ages three, five and six. The two youngest were mine by birth and the oldest I just claimed as my "extra" daughter. She was a perfect stair step for my two. No amount of looking would show anything to say that she didn't belong to me. Besides, I wished she were mine. Her Mother, Molly, was out in the rain, building my fence.

The previous autumn, when Molly had no firewood, I had given her a load of mine and then cut her a good bit more. The trade was that she would build my fence if I provided the materials. Then she ran out of firewood again, but wouldn't ask for more until the debt was paid. On the second day, when the rain had turned to sleet, the fence was about half finished. But I figured that she had already paid more than her share. I always figure that one person's day is the same value as any other person's day. It only took me half a day to cut the wood for her, so even if you counted the saw and truck as another half day, she had already paid. The problem was that Molly is stubborn.

I pulled on my gumboots, told the kids to be good and went out to the woodshed. I filled my wheelbarrow with a substantial load of wood—most of it good dry stuff—and then walked it over to her porch and dumped it there. Molly's house was located on my property. As I walked past her, I could see that she was pretty done in. I told her there was a cup of hot tea waiting for her inside the house. I knew that by the time she drank the tea and we talked a bit it would be dark and she wouldn't have to go back to the job again. Over tea I told her that I figured we were about even and that she should call it quits.

"No way," she said, "It's got to be finished."

The next day we bundled up the kids and set them to playing where we could keep an eye on them. We split rails and planted posts until we had it beat. The day after that, we cut and split a truck load of wood together and shared it fifty-fifty. That way nobody owed anything.

As I said, Molly lived in a house on my property back then. My house is eighteen feet by twenty feet; Molly's was twelve by sixteen. Molly's place had been on my land, half-finished, when I bought it. It was supposed to be an artist's studio. When I bought the place the owner sold the half-finished structure to Molly and her old man, who lived a few hundred yards down the road. There were no cops on the Island in those days so we got a forklift and a couple of 4×4s and dragged it the three hundred yards to their place. It sat there just as we had left it for some years. Sub-floor, stud walls and a shake roof. Once in a while when there was no kindling for the wood stove Molly would go out and split a shake or two off the roof—but only from the edges of the roof.

When Molly and her old man split up he stood in the doorway and screamed at her that she and her daughter would freeze in the dark without him. This was really a joke of some sort. It was Molly who provided all the firewood and what little food there was in the house. The bad part was that Molly believed him. She showed up at my place figuring that she and her daughter were going to die a slow and miserable death.

We talked things over for a bit and then made a deal. Molly informed her ex-old man that the unfinished artist studio was hers, and she was taking it with her. I happened to have a Cat working at my place clearing land, so we sent it off to get the studio. In less than an hour the Cat was back with the building and we got to work. The deal we made was that I was to supply the materials and finish the house to the point that it could be lived in. Molly was to have two years rent free, then it would become mine and we would negotiate the rent thereafter.

In those days we all had wood heat and kerosene lamps. Water was in buckets from the well. Everything was simple, so we kept it that way. Besides, we didn't have money to do it any other way. The walls went up and a chimney and barrel stove were installed. A double bed was built into one corner with a small loft overhead. A sink and counters were built on either side of the stove. It was too far from Molly's house to the outhouse we used, so we found some scrap lumber to build a nice one close to her place. It looked out over the beach.

When Molly moved into her new place the right and proper folk of the town were horrified. Molly had a few habits that kept the proper folk in something of an uproar. Molly was very pretty, and she loved alcohol, good times and men. A lot of the townspeople thought of her as down-and-out and said she should change her ways and apologize for being the way she was. Since Molly was just fine with the way she was she didn't do any of the above and kept on exactly as she wished. Molly had more than her share of rough times—most brought on by herself—but she was never down-and-out. And she was never a loser.

Molly had a kind of integrity that they don't talk about in books very often. We lived about five miles out of town. When Molly's daughter needed to get to preschool Molly found a way. They would start walking with lots of time to spare so, if they didn't catch a ride with a neighbour, they would still get there. There wasn't much chance that they wouldn't get a ride, but sometimes there was just no traffic, and they always got there right when they were supposed to. They returned home the same way. The daughter, at least, was always dry and clean. There was one very religious family that wouldn't pick them up because "Molly and her daughter were the wrong kind of people." They likewise wouldn't speak to me because I associated with Molly and her daughter. A few years later the religious family fell on hard times and had to ask us for rides almost every day. We never said a thing, but it was easy to see that they remembered.

The way Molly lived upset the social workers a bit. University-trained people can never seem to get over the idea that a house has to be full of stuff in order to be a home. Molly's house didn't have much in it, but she and her daughter spent hours a day together on the beach. Her house was only fifty feet from the high tide line. They played, watched the birds and animals and collected firewood. The social workers thought the daughter should be in a foster home, or at least in daycare. There were certainly problems at times, but Molly's first instinct was to protect her daughter, regardless of the cost to herself.

I was lucky enough to have a bit of an "in" with the local authorities, so we made a deal. I could be an unofficial guardian at times, and if things went really bad the daughter could move in with my family. When Molly was too busy or too involved in what she was doing to provide a good home, the daughter came to my place, about a hundred feet away, and life went on as usual.

One of Molly's boyfriends was a one-legged, drug-dealing alcoholic who periodically took shots at the cops. She threw him out when things had plainly gone too far, and he went off to jail. One way or another, he died there. Molly got a package in the mail one day. It was Long John, or at least his ashes. There was a letter attached, informing her that none of his family wanted him back and that Long John had asked he be sent to her "because she was the nicest person he had ever known." He got one hell of a wake, a few tears, and a good bit of respect. Molly said that no one, no matter what, should be so lonely they couldn't find someone to spread their ashes— and to care. The fact that the ashes were used to fertilize a pot-growing operation was just poetic justice.

I helped Molly buy her first chainsaw. In those days, in that place, a chainsaw was a meal ticket. There was always someone with some money who needed firewood, or some land clearing done. If you also had a truck, you had it made. Molly's theory was that so long as you had the best chainsaw you could afford, the truck didn't matter. She would scan the papers for "junkers'" and then offer half the asking price. Some of them ran for ages while others just died at the side of the road. When the trucks died, Molly just walked away. If she was near home, she would have a friend tow it to the car dump. If she was out on the highway she would just take off the plates and registration and stick out her thumb. The main trick was to never haul more tools than you could carry away in your arms.

After a year or so of odd jobs, Molly got an offer too good to miss and moved to another town. While she was moving a policeman stopped her on the highway for not having tail lights. After that she had to get a driver's licence, too.

Molly's life was no easy row to hoe, but she had friends who had it a good deal worse. There never is an adequate explanation as to how a person gets into some of the situations they manage to become involved in, but if you called Molly for help you could count on her being there—no questions asked.

Monica called Molly one afternoon and just whispered, "Help." Molly already knew half of that story and was afraid she could guess the rest. Monica had a boyfriend in jail. Monica also had a boyfriend not in jail. Monica was also pregnant, but only she and Molly knew that. The problem came in the form of a phone call from the bus station. Boyfriend number one was unexpectedly out on parole and would be home in a few minutes. Neither boyfriend knew anything about the other. By the time Molly got there, the two boyfriends were sitting across the room from each other. One was holding a 12-gauge shotgun, and the other was holding a .357 pistol. They were both drinking whisky. Monica and her infant son were sitting at the kitchen table. Monica kept making appeals for sanity and understanding, but no one was listening.

Both boyfriends were of the opinion that Monica had created the problem and had to be held responsible. They also both agreed that the real problem was the person sitting across from himself. The house was an old floathouse, just a cabin really, and it had only one door. For Monica to get to the door, she would have to walk between the two men, each of whom was holding the other at bay with his loaded and cocked weapon.

Molly walked in the front door without knocking, but neither man took his eyes off the other. Molly gave a big smile and said, "Whoa." Like Moses parting the Red Sea, she walked between the men and the guns and went straight to Monica. "Time for the women and children to get out of the way," said Molly. She picked up the baby, took Monica by the hand and led her between the men and out the door. "You guys just take it easy for a while, we'll go get some pizza."

My house is thirty-five miles and a thirty-minute ferry ride away from Monica's house. Around bedtime a car pulled into my yard and drove right past the house and up into the bush behind. I watched with some curiosity as Molly, Monica and the baby walked back out of the bush. They announced they would be staying for a few days, and that it would be best if absolutely no one knew they were there. We decided that a narrow escape had been had by all, and that a party was in order. Monica was both nursing her son and pregnant, so she drank pop. Molly decided that she should do the honours for the both of them. I had a couple of glasses of wine. Molly had a bottle of brandy, and after that she managed to get to sleep. Monica is one of the more beautiful women in the world, so after Molly passed out I tried to convince Monica that she should allow me to adore her properly. She said that being held at bay by two armed and angry drunks had put her out of the mood for such things. I had to make do with dreams of what might have been. A couple of days later Molly, Monica and her baby headed south to the women's clinic, where they solved one of Monica's problems. At some point someone told the cops that the parolee boyfriend was drinking up a storm— in violation of his parole. He was escorted back to jail. Monica threw the other one out. After that her life got much simpler.

There's no end to stories about Molly saving someone's tail by performing some unbelievable heroic act. There's even a rumour that she once robbed a store to get money for a friend who had to get away from a brutal husband. I asked her about that once. She figured that if it didn't happen exactly like that, it could have, so it was really all the same.

Molly has integrity. I've always said that if I found out I had to go to hell tomorrow, Molly would volunteer to come along to hold my hand and keep me from being afraid.

Happy endings are rare things, and like they say it ain't over till it's over. But things in Molly's life have been working out pretty well for a while now. She gave up booze. She became a mother for the second time, gave up men and became a grandma. Molly was always one of the smartest people I knew, so it was only a little surprising when she got a job looking after people who live like she used to. It's not easy at times, and there's no shortage of problems, but she's making it in fine style. And just like she always did, she's doing it with love and integrity. Life never does turn out like you plan, and sometimes things come out right no matter how hard you try to mess them up. But I still wish I didn't have to give the daughter back.

WAITING FOR THE TIDE TO TURN

Some people are waiting for their ship to come in. Others are waiting for the second coming of whoever it is they hope to hang around with for the rest of eternity. Bob Dylan is waiting for Quinn the Eskimo. And I'm waiting for the tide to turn.

It's very nice to have a panacea in your life—some future moment that will cure all ills, make you happy, and fix your car. It seems to me that the sooner it arrives the better I'll like it, but there seems to be some kind of inverse ratio operating in this regard. The way it comes across to me is that the bigger the fix, the longer you have to wait for it. Jesus, Buddha, Muhammad, and all the other prophets and avatars will fix absolutely everything and say, "I told you so" to the losers. But you have to wait a long, long time for it to happen.

Ships do come in, but they are notoriously slow and have a nasty habit of having cargo onboard that you hadn't planned on. Quinn may or may not show up—it all depends on the walrus hunt next year. But the tide turns every six hours, and there ain't nobody can deny or stop it.

Now, according to the inverse law of "wait time versus legitimate expectation," you really can't have that much hope for an event that occurs every six hours. But I say, "what the hell, go for it, hope for the moon." If you hope often enough for enough things, you're bound to get lucky sometime, and things will come true.

I checked with some of my fishing friends to see what kind of results can be expected from a tide change. According to expert opinion, it goes something like this: If you are a cod fisherman the wind will drop off. If you're a gillnetter it will come up. If you're a troller it will get better. If you're a seiner, who cares—you're tougher than wind. Fish will likewise do the right thing: gill, bunch up, rise and start biting. Except for dogfish, which will not gill, will go down, spread out and not bite. Unless you're deliberately fishing dogfish, in which case see list one above.

What it all comes down to is that everyone is just waiting for a change in the status quo. It is in this regard that I make my argument for the turning of the tide as the best symbol of hope. If you have to wait two thousand years for the big change you can afford to be pretty smug, but you've still got all your luggage in one cart, and if it fails—well, it's a big let-down. Also, if you're waiting for an ill-defined event or cir-cumstance (e.g. your ship coming in) you have to be careful not to designate any particular circumstance as the expected arrival because if you say "this is it," then any subsequent fail-ure of the promise is YOUR responsibility.

But the tide turns four times a day. We even have little books issued by the government that tell us where, when and how much. You can even predict the tide in places where there is no tide, like Denver or Edmonton, and every scientist on the planet will swear that the calculations are correct and that the event does occur. The fact that there is no water there to be affected is irrelevant. An interesting technical aside; you don't have to be a rocket scientist to figure out the tides in the prairies or on mountaintops, all you need is a bucket of live oysters. Live oysters in Winnipeg or Kansas City will open when it would be high tide there if there were water there to have it in—which there isn't—but we know they still have tide because the oysters know it and their brains are too small and squishy to cheat, guess or lie.

Back to my theory. Having the turn of the tide as your causal force in life gives you a clean sweep every six hours. If the tide fails to do its duty as expected it's not your fault. At worst, you were guilty of nothing more than picking the wrong tide—probably it will be the next one, and if it's not, well then El Nino probably screwed it up.

So I sit, waiting for the tide to turn, getting ready for the way it will be, and hoping that I've got it right for a change. But if, as so often happens, I've based all my hopes and preparations on the previous tide, instead of the next one—well, there will be more after that, and if I hang on long enough, the one I prepared for will come around again, and at least once I'll get it right. Every six hours, all bets are off—and you get to start over again.

So fisherman John was a fellow who did everything right. He consulted the tide tables regularly, kept his boat in good repair and, like most fishermen, dreamed of a better day. John did okay but he never really hit it big. It wasn't so much the money that he wanted; it was the satisfaction he thought he might feel being the "high boat," not just for a day, but for a whole opening. It seemed like it was never going to happen, so John took some drastic measures to make it happen just once.

John never was much of a fisherman. To begin with, he didn't like killing the fish, and beyond that he just wasn't very good at it. But he loved boats; that's how he got into fishing in the first place. It was the only way he could figure out to hang around on boats and get paid for it. He had started out in yachts and seemed to have a promising career in that department. But eventually he discovered that the owners of the yachts didn't care at all about the boats that he loved; they cared only for the social status that owning a particular vessel brought them. To John this was sacrilege, and he moved on to trolling.

Commercial fishermen care little about their boats except as a platform to catch fish from, but they don't put on airs about it. Some have nice boats, some don't. You're free to do it any way that pleases you, and since you can justify almost any expenditure of time, effort and money as an attempt to catch more fish, no one asks you why you are spending all your time, effort and money making your boat the way you want it. This suited John just fine. He kept his boat clean, decorated it tastefully (as befitted a fishboat) and bought the very best of everything to help him navigate and catch fish. If you asked him, John would quietly admit that he took a great deal more pleasure in getting ready to fish than he did in the fishing itself. He loved arranging his gear and designing new bits of equipment that made the system work better. He was also something of a traditionalist, and often used designs he found in history books or old photographs. Often as not they worked just as well as the modern equipment, but gave John much more pleasure to build.

John got into trolling at a time when it was relatively easy to make a living doing it. The glory days were already past and the handwriting was on the wall for all to see, but still it was, all in all, not a bad way to make a modest living. John was never that interested in money, so as long as he made enough to live in his quiet way, he was content.

John's boat was an old but good design, built of the best timber that had been available fifty years before. She was deep and narrow, which made her easy to push through the water, and she was ballasted like a sailboat to keep her decks looking at the sun and her keel looking at the bottom. She rolled a bit at times, but always snapped back with a quick easy motion. It was this way of rolling that made John fall in love with her. He had been looking at modern fibreglass boats with semi-displacement hulls, but when he stepped on board they were as solid as a rock and scarcely responded to his presence at all. When he stepped on to the boat that was to become his life, she gave a little curtsy and responded to his every step as he moved about the deck. "She's alive!" he thought. Alive and aware—a good partner to dance with.

John took pride in the way the boat looked, and it hurt him to the core the first few years that he couldn't afford to paint her properly. Fishing was failing. Insane government policies, combined with the uninformed opinions of a few bureaucrats, had destroyed what over-logging and industrial pollution had left behind. When he started out, John worked ten months a year, and put in more than twice the hours that a government office worker does. By the end, he and his fellow survivors were begging to be allowed to work ten days a year—and mostly they didn't even get that. Thirty days a year would have been plenty. There was no shortage of fish and the price, while not as good as it had been in the past, was okay.

But the media had led the public to believe that there was a shortage of fish. There were indeed stocks that had been horribly abused and were on the verge of extinction, but overall the stocks were in great shape. The salmon were so thick that the department of fisheries had to hire boats to seine them up, kill them and dump them back into the sea, to prevent them from plugging the spawning grounds. Electric fences had to be put across the rivers and millions of sockeye bulldozed out of the water and hauled to dump sites. One year alone, twenty-two million sockeye, worth twenty dollars each, were dredged out of the rivers and buried—while the central coast fishermen starved.

John survived it all—just. To qualify for unemployment insurance you had to make at least $11,000 a year. That was $11,000 after a twenty-five percent reduction applied by the government to make it harder to get on "pogey." Most years John could just make it. He fished his salmon and jigged ling cod and sometimes paid taxes on money he didn't really earn, in order to appear to have enough income to qualify for the full benefit. The thing that John just couldn't do was be a high-liner and get the satisfaction of slinging more sockeye than his fishing partners and friends. Most of the other guys, at one time or another, had had a really great year or even just a really good trip, and out-fished everyone else in the area, but John had never done it.

Sockeye fishing is an underappreciated sport. You invest about $200,000 in a boat, licences and equipment, then spend untold hours dragging your trolling gear (about 108 lures at a time) through empty water. And then—suddenly—the gear starts shaking, the bells start ringing, and you start pulling in fish as fast as you can go. The fish come in at the rate of about one every thirty seconds on each side of the boat. That's about 240 per hour. Each fish is worth about twenty dollars. I once spent four consecutive hours slinging sockeye aboard at that rate. My partner that day was my daughter; she earned her whole year at university in one long day, and we laughed ourselves to exhaustion and kept slinging fish.

John had seen it often enough, and he had had the experience himself, but only for a few minutes at a time. John didn't want more money—though he needed it—he wanted the experience. He wanted to feel the exhilaration, and see the look of envy in the other fishermen's eyes, but as the fishery kept winding down John became convinced that it would never happen to him. John was used to the idea that life didn't necessarily work out the way he wanted it to. Two failed marriages and a child who wouldn't speak to him had led him to be philosophically reconciled. But the bad choices and hard luck hadn't extinguished his innate hopefulness, nor quieted his bright imagination and sense of humour. John decided that if he couldn't have the real experience, he would fake it.

John had a grandson who loved him without limit, and because John was a fisherman, the boy loved fish. Sometimes when he was visiting at his grandpa's house he would ask that one of the many fish in the freezer be brought out so he could play with it. John kept one special fish for the boy to play with. Over time it became worn and freezer burned, but it never failed to excite the boy's interest, and he always ended the visit by begging to go fishing with grandpa. John always promised that they would go fishing together, though he secretly feared that by the time the boy was old enough, there would be no fishery to go to.

Whenever John had to go to the city—usually for a doctor's appointment—he would go to his favourite toy store, the Red Snapper, and buy a small present for the boy. On one such trip to the store he discovered some small rubber fish. They were poorly designed and only about eight inches long, but he bought them and they planted the seeds of his plan. The lady who owned the store gave John the name of her supplier and from there he got into contact with the factory in Taiwan that made the rubber fish.

The negotiations for the order that John wished to place were not so much difficult as they were absurd. Despite the fact that the agent he spoke to used perfect English, there was poor communication. John sent drawings and photographs and sketches with exact measurements of what he wanted. What he got in return looked like something left over from the set of "Godzilla Meets the King of the Piranhas." In the end he settled for something that could be mistaken for a sockeye once he painted over the bright orange fangs. John only needed one-hundred-and-eight of these monsters to put his plan into effect, but the factory insisted that the minimum order was one thousand. John had to dip into his very meagre savings to pay for the order—plus freight.

There are areas on the coast where boats fishing for sockeye are spread miles apart and seldom see one another. There are other places where, at any given time, there are a hundred boats locked in a tight intertwining knot that scarcely covers an acre of water. The most notorious of these is in Robson Bight and it is known (in its polite form) simply as "the circle."

Imagine a hundred trollers, each dragging at least a hundred lures, packed so tightly together that they are only a few feet apart. The bow of each boat is right up between the lines of the boat ahead, and the tips of the trolling poles of the boats on either side are barely feet apart. Now add vast numbers of fish, and levels of testosterone, adrenalin and rude machismo that no football game could match. There's a fortune in the water, a gold rush—and it's all in one place. If one boat gets a school following it and tries to escape from the circle, the other boats will follow it and box it in until the school is all caught. Sometimes the school won't know which boat it is following and everyone will be pulling fish as fast as they can go. No matter how fast that is, the skipper will be screaming at the crew to work faster, and at the boats around him to back off, or speed up, or slow down. It's a feeding frenzy, and many good skippers refuse to participate. Some skippers, however, live for these moments. These are the same kind of men who intentionally go to a football game where they think there is likely to be a riot. They reinforce their boats and gear and plan on collisions and entanglements. Their hope is to ruin the other boat's gear while continuing to fish themselves.

John never went to the circle. When he was younger he had tried it a few times, but not found it to his liking. Like many other trollers, John thought of himself as a gentleman engaged in an honourable pursuit. His life did not include temper tantrums and screaming obscenities at his fellow-trollers. But this time things were different.

When John went to put his plan into effect, he headed right for the centre of the circle. There is actually a circle within the circle, and it is the home of the hardest of the hard core skippers. If you limit yourself to the outer circle, you can bail out when things get too crazy, or just ride on the outer edge getting bumped farther and farther out as more boats join in. If you make your way to the inner circle you are there for the duration and the rules are cold and tough. Big steel boats go where they wish, and move as fast or slow as they wish. Smaller, more manoeuvrable boats dance between the big ones without regard to right of way, or to the disasters that occur to the other boats that try to miss them. The skippers don't look back and they don't negotiate position on the radio. John went for the middle of the circle.

A working troller has to put its lines up and down on a regular basis. You throw out the gear empty and pull it in when you think it's full of fish. John reversed the process. As his lines went over the side, he quietly hooked a big rubber sockeye on each and every hook, one-hundred-and-eight in all, and then threw a big heap of several hundred of them in the checkers, where the fresh fish go. Then he forced his way to the heart of the circle. Once he was firmly established in the holy of holies, he started pulling his lines. All hell broke loose. It had been a slow morning. None of the boats had more than a handful of fish on board, and many had none at all. When John started pulling fish as fast as he could go, every skipper there tried to get as close to him as they could. If you cut close to a troller that has a big school of fish following it, some of the fish will switch over to your boat, or in some cases you can steal the whole school. The fleet closed on John like a steel trap. He was locked into the centre of a feeding frenzy with nowhere to go. In fact, the fleet that was so desperately trying to get his fish formed a protective barrier around him.

And then he did the unthinkable—he went straight. From the dead centre of the circle he put the autopilot on straight ahead, turned off his radios, and walked back to the cockpit.

I was half a mile away on a much more civilized tack when I heard the radios explode. Screams, howls and obscenities of the first order—looking through the binoculars it was a scene of madness. Boats were peeling off in every direction. Poles were clashing, gear was tangling, and coming out—straight out—of the heart of the circle was a tight knot of boats moving like a flying wedge through the chaos. By the time John got to the outer edge of the circle he had pulled all his rubber sockeye and had put his gear back down empty—as it should be. There is a standard procedure in sockeye fishing where you do not start pulling fish as soon as you get them on the hooks. You let them build up there to form your own captive school, which lures other fish to join in.

As John put his empty gear back in the water it began to fill up with real fish, and by the time he had pulled all his rubber fish, his lines were full of real sockeye and he had to pull his lines again. And they just kept coming. Once he was well clear of the circle, but still surrounded by the ravening hordes, he made a hard left turn and headed west. It's a legendary move. If you get the mother of all bites going, head west and don't stop until you have the whole school in your hold. John kept going through Glory Hole, Izumi, Hillies Camp, Blinkhorn, Telegraph Cove and finally Lewis point. Beyond Lewis Point it was closed for commercial fishing. But John didn't care. The checkers were full to overflowing. Fish had spilled over and covered the deck, and as near as he could tell he still had a fish on every hook. He took a long slow right turn that put him on the Hanson Island side of the Strait, and trolled east. There were still fish with him.

By the time John got to Blinkhorn Light, there were only three boats still following him and by the time he reached the Hanson Island shore he was alone. It was a good opening, the best he ever had. He heard later that the boats that hadn't followed him had hit the fish a short while later and the chaos of the circle had re-established itself in its usual form. When he got home, John gave his grandson two of the rubber sockeye and at other times sent them as presents to all the little kids in town when they had a birthday. It wasn't all that long before they started showing up in the local thrift store and years later you could usually find one there, hidden in some dark corner.

THE SKIPPER AND
THE STRIPPER

A couple of the boys got tired of killing fish and trees for a living, so they decided to go south and make their fortune pushing big boats around with little ones. They convinced a fellow who owned a couple of tugboats to let them run them, and they figured they were sitting pretty. The contracts they got were pretty cushy. The tugs tied up in town every night and often enough half the day.

Every once in a while they would get a call and quickly run up to the pulp mill to push a big freighter in or pull it out, but there wasn't really much to it. Sometimes they were called to help a few barges through the rapids against the tide. That was more interesting. Sometimes a "wild one" would take on a mind of its own and slide all over the narrow channel, threatening by turns to run itself up on the rocks or to wipe out all the traffic coming downstream with the tide. Things could get messy and dangerous in a hurry. The problem was that the narrow channel was just a little too long for a big tug and tow to get through on one slack tide and they always ended up blowing black smoke, making less than a knot uphill against the current.

In these situations the current was always in the process of getting stronger and wilder, and the tugs were in the process of going slower and slower. Sometimes they lost the battle and had to back off the throttles and let the current push them back to where they had started. They would still be facing upstream, pushing against the current, but they would be going backwards through the narrow channel. In a situation like that a person could start feeling real old in a big hurry.

When the boys got back to town it was only natural to want to relax a bit and sip a few cool ones. There were two watering holes in the immediate area. The nicer one was the home of a biker gang, and though they didn't mind tugboaters drinking in their place, they liked them better if they rode bikes. These particular tugboaters didn't ride, so they went to the other place.

The watering hole of choice was also a notorious strip club. Most tugboaters have no problem relaxing with a few pints between jobs, while watching an attractive young lady remove her clothing and strike various poses designed to make you think she has the hots for you—just you, and not the other thirty or so guys in the room.

One of the guys watched the girls strip and talked in a quiet way about the previous job or upcoming repairs on the boat. The other guy had a problem. He really believed that the stripper he watched was dancing for him, and him alone, and that she was making promises with her eyes. When the time came to go to the next job, the fellow was so horny he could hardly walk, and he felt cheated. He felt so cheated he got angry and nasty. Once he was angry he wanted to talk about how badly he had been treated.

As the tugs ran to their next appointment he would talk on the radio the whole time. Often enough he was quite graphic about what he wanted to do to the little minxes that kept teasing him and not keeping their promises. It was so bad that the other guys got in the habit of turning off their radios and not turning them on again until they were right up to the freighter they had to move. This could easily have resulted in any number of problems, but for the most part things went well.

The skipper with the stripper problem even started going to Zellers and buying women's sexy thong underwear. These were pinned up in his tug's wheelhouse and he would tell anyone willing to listen exactly what he had done with the girl that used to wear them. To hear him talk you would have had to believe that he was the greatest stud anywhere and that women sought him out day and night.

The crew of the other tug also took a trip to Zellers and bought a pair of women's underwear that they nailed up next to the thongs on display. This pair, however, was somewhat larger than the others—size double-extra-large, with pink polka dots. The skipper was enraged and had to rip them down because he couldn't find a hammer to pull the nails.

Due to the woman-crazy skipper's attitude his own crew, as well as the other crews, started avoiding him in the pub. Every time he showed up, everyone else was just on their way out. One day when the crazy skipper wasn't there, the other boys were sitting sipping and watching a stripper.

One of them mentioned that he was a good friend of the stripper then on stage. To make his point, he waved at her. The stripper waved back and gave him a big smile. When her act was over she came down and sat with them. Introductions were made and everyone agreed that it was too bad the woman-crazy skipper wasn't there. They figured that he would have gone nuts being so close to the object of his desire.

Someone suggested that, since the crazy skipper wasn't there, they should take the stripper to the skipper. A plan was hatched and everyone, including the stripper, left the club to put it into production. The tugs were due to meet in about an hour to push a big freighter into the pulp mill, and they would be operating almost side by side. The stripper had to be back for her afternoon show, but everyone promised her that she would be there on time. To make sure she made it, one of the fellows drove his truck up to the mill and parked on the end of the pier.

The stripper was stashed below deck on the first tug and given a cup of coffee. The tug pulled out, even before the crazy skipper had got down to his own boat. When he saw that the other tug was already on its way out he called them on the radio, but as usual nobody answered him. His crew, being in on the whole deal, just yawned and played dumb. The second tug followed the first one out and they waited for the freighter to get into position.

Once things were lined up, the tugs moved in and started leaning on the freighter. There was nothing especially difficult about the job until the crazy skipper heard loud music coming from the deck speakers of the other tug, and looked to see what was happening.

The tug boat with the stripper on board had an old-fashioned cabin arrangement. The wheelhouse was right forward with a long flat-topped trunk cabin aft that covered the galley and crew's quarters. The flat roof made a perfect stage, the deck speakers were blaring her music, and the stripper was there doing her full act.

The crazy skipper went nuts. He started screaming into

the radio, demanding to know what was going on—no one answered him, and the girl kept taking it all off.

There is nothing especially difficult about docking a freighter, but you do have to pay attention. The crazy skipper had eyes only for the girl, but he had to keep looking away to pay attention to the job at hand.

The girl was in on the joke and pulled out all the stops. The crew on the freighter heard the music, and so they congregated on the rail to watch the show. They were cheering and yelling encouragement to the stripper, but she had eyes only for the distraught skipper in the other tug, and made any number of promises with her eyes—and other portions of her anatomy.

The tug with the stripper on it backed off the freighter the second it touched the pier, and spun around to the other side of the dock. The girl jumped ashore, ran to the truck that was waiting for her and was on her way laughing.

The other tug had to hold the freighter against the pier while the lines were made fast. But it was still less than ten minutes later that it arrived alongside the first tug. The crazy skipper jumped on board, frothing at the mouth. The other skipper and his crew looked up from where they were quietly having coffee at the galley table, and nobody said a word. The crazy skipper looked around wildly, then ran through the whole boat searching for the girl that he was sure he had seen stripping on the roof. No one else had seen her. Even his own crew, who came in after they had finished tying up, didn't recall ever seeing anything out of the ordinary.

A QUESTION OF SMARTS

Some questions are better not asked. Not because the answer is difficult, or hard to find, but rather because the answer is just waiting, begging, to leap out at you. And you know that you're not going to like it at all. Occasionally, someone will phrase a query in such a way that, whereas the question remains implicit, it is obviously removed even from rhetorical commentary.

In my life, most such statements have to do with the concept of "outsmarting fish." See! You already know what I mean and just what the problem is. Fish have brains the size of a pea. They're not smart enough to know that worms are a disgusting breakfast food, and yet each year millions of would-be fishermen spend countless hours and millions of dollars trying to outsmart fish.

I have known men of high professional standing, lawyers and accountants, men respected in their fields, who have spent weeks planning their assault on some poor dumb fish that was in all likelihood no more than a rumour, invented by an overly enthusiastic travel-brochure writer thirty years ago. And yet, not one of these men was the slightest bit ashamed or embarrassed by entering into this contest of intellects, even though deep in their hearts they knew they were almost certain to lose.

Outsmarting fish involves three basic steps.

1. Getting to the fish. This usually requires a 4×4 pickup truck with oversized tires, a roll bar and racing stripes, combined with a colour-coordinated boat, featuring twin 200 hp outboards on the back and great big brand name lettering on the side.

2. Attracting and hooking the fish. This is the crux of the matter. Whereas everyone knows that fish eat amazingly cheap food like worms-and-herring, most fishermen, after a great deal of research and discussion, opt for amazingly expensive neon-plastic and chrome alternatives. We of the worms-and-herring school of thought believe that the neon-and-chrome crowd are giving the fish more of a sporting chance, by luring them with something they wouldn't normally think of eating.

3. Landing the fish. Commercial fishermen, who land lots and lots of great big fish, use heavy lines and hydraulic line pullers. Other fishermen, hoping to catch fewer but even bigger fish, use extremely light carbon-fibre rods and manual reels made in the Japanese equivalent of a Swiss watch factory. We don't know why, but after a fisherman gets serious and thinks about it for a while, it seems to be the thing to do.

The thing about the whole system that becomes clear is that—whereas some fishermen catch some fish some of the time—most fishermen appear to be themselves caught by a huge conspiracy that lures and catches fishermen. The obvious question then becomes, are these anglers of anglers themselves, sought and caught by yet other anglers angling for them; and if so, where does it all end?

It may just be possible that the entire economy of the Western world depends on the atavistic desire that men (and a few good women) have to stalk and capture a few intellectually challenged fish.

As for the fish themselves, they appear to be in no significant danger. For while there are millions of people spending millions upon millions of dollars and hours upon hours of time trying to outsmart them, the thinkers have yet to succeed, and the fish with their pea-sized brains are just too dumb to know or care.

We use herring for bait. Live herring if possible, but pickled or salted will do if that's all there is. We have various schemes for catching them. Sometimes we get a whole fleet together. One boat carries the bait seine. The others hunt the herring and stand by to pack it in live tanks for transport back to the holding pond. We take turns working on the seine boat. There are usually three of us, but two will do in a pinch.

It was one of those days where there wouldn't be enough packers if we took a third man on deck, so there was just the skipper and me. Being short-handed, I laid out everything nice and neat so there wouldn't be any screw-ups. I put the sea anchor on the afterdeck and put the scotchman on top of it. All ready to go.

To get where we wanted to go, we had to go through the Blow Hole. There are several places on the coast that have this same name. What they all have in common is that they are very narrow and the current runs very fast. I guess the current made us roll a bit. At any rate, the scotchman went over the side and took the sea anchor with it. Next thing I knew, we were setting our net right down the middle of the channel and the tide was running like a river.

For you non-seine people, just think of this as an experience that you would really rather avoid. Something akin to spilling five gallons of used crankcase oil on the middle of your mother-in-law's new carpet, just before her bridge club arrived. Strangely enough I have seen this same kind of thing happen in town. One time, a seine skipper was moving his salmon seine down to the harbour when the end fell off the truck. By the time the driver saw what was happening, there wasn't enough left on the truck to worry about.

I was new to fishing at that time, and I couldn't quite figure what the point of the operation was, or why it took so many people to comment on it. The net I dumped in the Blow Hole didn't go nearly that far, but it was all that the two of us could do to get it back on board. The other boats were well ahead of us, so no one saw my disgrace, though of course the skipper told the whole bloody coast as soon as he got near a radio.

The easiest way to get herring is with a hand seine. There is a certain place in the Gulf Islands where the lights of a pub draw the herring right up to the loading dock. The floats at the bottom of the ramp are perfectly positioned for hand seining, and there is beer at the top of the ramp. We stretch the net between the floats, then adjourn to the pub. After what seems like the right amount of time, we signal the barkeeper and she turns off the dock lights. Long experience and much careful experimentation has shown that there is exactly enough time to have one more beer, pay the tab, and slide down the ramp. When the lights go out the herring start slowly rising. If our timing is correct they will arrive at the surface just as we arrive from the pub. We will pull in the seine, brail them into a holding pond, and go to bed. In the morning it's business as usual.

The way we usually get herring is somewhat different. Herring are schooling fish. Eons of evolution have taught them that their best chance of survival is to stay in schools, and to get closer and closer as the predators close in. The result is what we call a "herring ball." Pursued by salmon, seals, dogfish and diving ducks, the herring swim into a tight knot just below the surface. As the pressure from below increases, the fish on the top are lifted right out of the water by the fish underneath. Sometimes there is a mound of herring eight inches high, flipping and jumping off the ones below, trying to get back in the water.

Somewhere in this cycle the seagulls spot the rising ball and go hysterical with anticipation and delight. Ducks can dive thirty fathoms with ease, but seagulls can only reach down a few inches. Hard as they may try they are limited to the depth that they can reach with their backs still above the water. So they have to feed on the surface.

The gulls form a distinctive spiral of diving, screaming birds, and that's what attracts us. The balls can last seconds or hours. Sometimes the ball dives and we can see it on the sounder far beyond reach; sometimes the whole thing is devoured long before we get there, leaving only a cloud of scales in the water to show where it had been.

When we see a big group of birds screaming and diving it's time to get the hoop net out. My hoop nets are made from a heavy steel ring about three feet in diameter and a long sock-shaped bag of web trailing behind. You run up to the herring and throw the hoop into the ball. The heavy ring drags it down, trapping the herring, but only so long as it is moving. As soon as it reaches the end of the tie-up rope—don't forget to secure the tie-up line—the net stops sinking and the herring can swim out. So the crew—or skipper, depending on the set-up—start pulling the net up as soon as the boat stops. There is an arrangement on the side of the boat to hang the hoop so that the hoop is open at the top and the herring can be scooped out (brailed) and put in live tanks for transport back to the camp, where they are transferred into a holding pond.

We see a ball starting up. We turn our boats toward the screaming gulls, and put the throttle to the wall. The crew goes forward to get the hoop net ready to throw. In the last few yards, the helmsman studies the birds to find the exact spot to throw the hoop. As the boat hits the cyclone of birds, they explode in all directions screaming their outrage. At the last moment you pull back on the throttle and glide past the spot where you hope the ball is. The hoop is thrown, then it's hard over with the helm to kick the stern away from the net, and reverse to slow (but not stop) the boat. If you stop completely, the herring can get away. All in all it's a lot like calf roping at the rodeo, and just as much fun. What with cross-tides, and junk in the water, the number of things that can go wrong seems endless, but somehow it's still the best way to catch herring.

Sometimes there's competition for the balls. Minke whales love to eat them. Sometimes they run on the surface and dive down on the ball. Other times they come up from below, taking in herring, dogfish and gulls in a single slurp. A friend of mine went to throw his hoop, only to have a Minke come up through the ball at exactly the same moment. He stood three feet away watching the herring, dogfish and gulls go down the giant maw. Then he turned and said, "That was the end of their whole little universe." Then we tried to beat the Minke to the next ball. The whale got three in a row; we got number four.

In the fog we can't see the seagulls, though we can often hear them. Mostly we just drift in the fog, and when a seagull goes by in unseemly haste we follow it. Forty feet of boat, 120 horsepower, two radars, two sounders, five radios, autopilot and GPS–all following sea gulls in the fog, hoping they know where they're going.

LUCKY

I once ballasted a boat with beer. It was a desperate situation. We were in the middle of the fishing season and word came over the radio that a beer strike was imminent.

The skipper had gone home for a couple of days, so I knew that I had to take decisive action. I headed out at once and got to Alert Bay before the word had spread too far. All my friends heard I was on a mercy mission and asked me to pick up a few flats for them too.

The boat was a lovely old thirty-five-foot troller with lots of room in the fish hold, so I started laying in the cases. When Alert Bay was out of Lucky, I headed for Sointula, and then to Port McNeill. By then I knew I had more than enough, but there was still cash in the coffee can where we kept the money for boat expenses, and besides, it was kind of fun.

I somehow convinced myself that I could replace the cash I was using and make some good money on the side bootlegging beer to my friends—if the strike went on long enough.

When the skipper got back he was pleased with my foresight but not terribly enthusiastic about the amount of money I had spent. He was also concerned about how we were going to tell my cases from his. He didn't mind donating me a beer once in a while, but he was damn well not going to subsidize my drinking habit. I was somewhat amused at this latter concern. I usually drank two beers a day, while the skipper drank twelve and then offered them freely to his friends in the evening. We solved the problem by putting one case off to the side, just for me.

There are many arts and crafts that started out as sailors' pastimes and have later been adopted by society as a whole. Fancy rope work, scrimshaw, and macramé, have all become known beyond their maritime origins. But one of the finest and most practised of the fishermen's arts remains almost completely unknown to the outside world. I refer to the fine art of "beer gaffing."

Beer is, of course, usually kept in the fish hold because it is cold and rolls about less down there. Although there is usually a small access hatch to facilitate getting into the hold, it is by nature an unpleasant place (wet, cold and slimy) and it is only natural to avoid entering it if possible. The wise fisherman always leaves a few cans or bottles directly below the access hatch, where they will be within easy reach of the gaff. If one is drinking canned beer, and one has likewise been clever enough to leave the plastic rings (so abused by the environmental advocates) in place, then it is a straightforward matter to reach down with the gaff and snag up some beer. This is not, however, considered a classy way of doing things, and will gain you no brownie points. A practised hand will casually reach down and hook a very sharp gaff under the bottle cap or outer lip of the can. This calls for a steady hand and is subject to two potential problems. First of all, the can or bottle may slip and fall back into the hatch, thus rendering it unfit to open for some time. Or secondly, you may accidentally pierce the can and have to drink the fine spray coming out the side.

A true aficionado, drinking from cans, catches the pull-tab with the point of the gaff and then, with a deft flick of his wrist, opens the can without having touched it with his hand. The finest practitioner of this art that I have ever had the honour to see in action could tip his head full back to drain the last drop from a can, while at the same time opening the hatch with one foot and gaffing a fresh beer, seemingly by instinct—sight unseen! (Yeah, Russell!)

We once tried to get the makers of Lucky Lager to provide us with fleet flags (our fish buyer was too cheap). We had an artist friend draw up a design—a rock cod rampant on a Lucky label—and sent it to them, asking them to make flags for us. We swore that we would call ourselves the Lucky fleet. The only thing we ever heard from them was some gobbledygook about patent infringement and something called a cease and desist order.

We all started drinking Molson after that. Eventually most of the guys figured out that you can get eighteen Luckys for the price of twelve Molsons, so they forgave the Lucky people and went back to the most for the least. It's a good philosophy and extremely appropriate to the wages we earn.

When the beer strike was over, the season was ending. The beer was gone, and I hadn't made a cent on it. I guess I just don't have what it takes to deal in the higher financial realms. In fact, the skipper never even paid me for all the beer he drank, and every time he was feeling magnanimous, he would offer me a beer that I had in fact paid for myself.

NEKKID WOMEN

Back in the '60s, it was quite possible to take a girl on a first date and go skinny-dipping. It gave both parties a chance to "check out the merchandise," but it did not necessarily mean that touching was allowed. Attitudes to nudity come and go, but outside of the city things tend to be a little more relaxed.

There used to be a small café at the end of Fisherman's Wharf. The takeout window faced the pier and it wasn't uncommon to walk in to find your coffee waiting and your usual lunch on the grill. Small town, just the way it should be. The owner-operator was a good-looking, well-endowed lady who was said to appreciate men as one of the finer art forms. I'm not spreading malicious rumours here—she is the one who said it.

So a couple of the local boys were picking up a nutritious meal of monster burgers and deep-fried bean sprouts—something like that—when one of them suggested that topless service at the window might generate a significant increase in tips. The mood being right, and the blouse she was wearing having a low and elastic neckline, she played it to the limit, and everyone laughed themselves silly, though one old-timer did look a bit startled. The resultant tip was indeed significant.

I was doing ten-day fish trips at that time. When I heard about the strip show, I suggested that it would be very nice indeed if on my return the first thing I saw when I came walking up the pier was a similar heartwarming sight.

Next time I came in, I saw her in the window and made the universally understood gesture, often translated as "Show me your tits!" I was at least fifty yards away and there was no one else in sight, so what the hell? That pink blob in the window, I was quite certain, was the top half of a woman. But really, at fifty yards away who could tell? The really funny part was that just as the show went on, the same old-timer stepped out from behind his truck and got an eyeful. As for being a "pink blob in the window," when the lady referred to read this particular adjective, she insisted that it be changed to blur, not blob, so as to put the fault on my eyesight, as opposed to any possible defamation of her figure.

Everyone knows stories about good and bad things happening in threes. Well, a short while later the talk in the café was about the aforementioned "topless show." One of the other local ladies was saying how, when she was out working in the logging camp, she had taken up nude sunbathing and lost all sense of modesty. She said it felt great to be free, and just to prove her point, she pulled off her shirt to show that she had a real good tan all over.

Really, it wasn't such a risqué thing to do—we were all old friends, and we were the only customers in the place. At least we thought we were. There was a startled gasp from the takeout window. We turned to see the old-timer's eyes getting bigger, and rounder, and bigger. He never said a thing, but a few minutes later when he picked up his order he stuck his head back in the window and said, "Thank you, thank you so very, very much!"

When I started working on seine boats, it was a common practice to idle slowly along the beach looking for schools of salmon. The normal procedure when fishing was to find a school and set a net around it. Every skipper had his favourite places to look. Most of the spots were traditional, so you could often see a string of different boats checking certain spots.

At some of these places we sometimes set our net and sometimes caught some fish. At other places we would just drift while the skipper watched the beach with binoculars. I noticed that the skipper spent long periods of time looking up on the hillside instead of at the water where the fish would be jumping.

"What's up there?" I asked.

"Nekkid women," he replied and handed me the binoculars. Sure enough there was a farmhouse with a big garden next to it. Inside the garden fence there were pink forms with long hair walking around weeding or harvesting vegetables. They may well have been humans with no clothing, but at better than a quarter of a mile in distance, who cares? I wasn't terribly impressed, but the skipper was. I watched the next boat behind us arrive on the spot. They too lifted their binoculars to the hillside. It was probably the safest place on the coast for a fish to hang out—nobody was ever going to see it there.

My own house is right on the beach and because of the way the brush grows along the road a part of the beach is quite private. Various young ladies who had a hankering for an all-over tan discovered that if they parked their cars at my house no one would suspect they were in fact down on the beach sunning themselves without the impediment of a bathing suit.

On the rare occasion that someone recognized one of the cars and wanted to talk to the owner, I would just walk over to the bushes and announce that so-and-so was looking for such-and-such.

It was a system that worked well and no one got excited, except for one poor fellow who just couldn't control himself at the thought of naked women on the beach. Every time he stopped he said he had to talk to whoever's car was in my yard. I usually informed him that she had loaned the car to someone else that day. I kept expecting him to make a break for the beach, and though I think he considered the idea seriously, he never actually did it.

I also had a large vegetable garden in my yard that was mostly hidden behind a tall deer fence. Some of the neighbours decided we should all garden together and share the proceeds. It seemed like a good idea to me, so on the first sunny spring day we all got together and started digging vegetable plots. It was a hot day in the spring, and so I eventually took my shirt off.

Next time I looked around everyone else had their shirts off too. There were in fact more women than men, but no one was concerned about such things. Well, that is, no one in the garden was concerned.

The telephone repairman happened to be driving by, and he must have looked at my garden through the one and only slit in the fence that would allow him to think he might have seen something he wanted to see again. He skidded to a halt and then backed up to the telephone pole in front of my house. The repairman put on his spurs and climbing belt, and ascended the pole. From the top of the pole he was in a perfect spot to look down into the garden. He didn't even pretend to repair anything. He just stared.

I presumed that he was staring at the women, not at me. The women also thought he was staring at them, so they started yelling abuse at him and holding their breasts up so he could stare at them better. "Ya wanna stare at my tits? Here, look at these."

The repairman finished "repairing" the line in record time and drove away. He was obviously embarrassed, but I noted that for several years after that he always looked to see what was happening in my garden. At least a couple of times he did get a lucky glimpse, but he never climbed the pole again.

I wouldn't want to give the impression that I was immune to the charms of such sights, but it didn't really affect me very much. I enjoyed it quietly, and took some pride in the idea that the ladies trusted me not to act inappropriately. That apparently meant "go ahead and enjoy the view, but don't stare, and don't say or touch anything."

A few years later I was running up to Port Hardy at the beginning of a fishing trip. My crew for that trip was an attractive lady who had been a good friend for many years.

Over time we had had hundreds of chances to get more friendly with one another, but we had always remained simply friends, so there was mutual respect, and modesty was not an issue. It's a long drag from Sointula to Hardy, and the day was hot and glassy calm. My crew asked whether I would mind if she took off her clothes and soaked up a few rays on the hatch cover. I told her it wouldn't bother me, but since I had to stay at the wheel I would have to look behind every once in a while to check for overtaking traffic. She said that was no problem.

She got some blankets from the forepeak and made herself a soft spot on the hatch. I looked back every once in a while, but I was far more interested in the traffic than I was in the naked woman on the hatch.

Sometimes, there's really no telling how a thing gets hatched in your brain. I think what happened was that I went to say something to the lady, but found she was asleep. So I just didn't say anything. What I had been going to tell her was that there was a passenger liner coming up close behind us, but she was asleep so I didn't say anything.

When the liner was right beside us and a fair number of people had gathered on the rail, many of them carrying cameras and video recorders, I stepped to the doorway and called her name. She opened her eyes to look at me, and I pointed up at the liner looming above us. It took a moment for her fuzzy vision to focus, and for her mind to acknowledge what she was seeing.

My plan had been to step back in the door and close it behind me, leaving her out on deck, but I never had a chance. She hissed a few unladylike words and flew through the doorway behind me. If I hadn't stepped out of the way I would have been run over. She stopped in the wheelhouse, wrapped a blanket around herself and suggested I try a few physiological impossibilities. Then she went below and got ready to start fishing.

As for myself, I couldn't see what all the fuss was about, but I did laugh a good bit about it.

WAR STORY

My first trolling skipper had spent World War II in the gumboot navy, alternately catching fish to feed the troops and patrolling for Japanese submarines. He said it was a good life most of the time, lots of hunting and fishing, and lots of time to explore every bay on the coast. He said he spent half his time in a skiff.

The problem was that they were always hunting for a rumoured Japanese sub that kept getting reported in the area. Since a sub would definitely outgun their patrol vessel, they had to take it by surprise. This frequently meant stopping on one side of a point and sending a man to the beach in the skiff, so he could look into the bay beyond to be sure the sub wasn't waiting in ambush. If the day was nice the recon would go on for some time. Naturally, it being wartime, the scout always took his .30-30 along (the fly rod was more difficult to explain). If the scout came upon a few deer or trout while reconnoitring, well, it was just natural to share his good luck with his shipmates.

There were, of course, black nights and horrible gales that threatened to put an end to all the good times. But those times being in the past, it was the good times scouting the bay that remained strong in his mind. Except for the damned "Jap sub."

Over time, they had interviewed dozens of good, reliable people—loggers and fishermen mostly—who swore that they had seen the damned thing. But the skipper and the crew never got a glimpse or whiff of it. They spent the whole war chasing an illusion, and thirty years later it still hurt.

At that time I had five acres of land and a house that was falling down. I decided to clear some land and build a new house. Toward that end I hired a local Cat driver. He brought in a friend. The friend was a school principal by profession, but loved to spend summers up the coast, anywhere away from the city. As we talked, it became apparent that he knew every nook and cranny on the inside coast. I asked him how, and he told me.

Most people don't know that during World War II, Canada had one submarine on the west coast. High authority decided that rather than send it off to sink the enemy, a duty from which it was very unlikely to return, it would better serve as a means of testing coastal defences.

"It was horrible," he said. We popped up in every bay and village on the whole coast anywhere there was a log boom or a shack, made all the noise we could; then we'd dive and run elsewhere. We'd cruise back and forth under the gillnet fleet at night with our deck lights on. No one ever saw us. We did everything possible to get reported, but as far as we know, not one single report was ever sent in.

I told each of the men about the other's story, but it didn't seem to make any difference. The futility and frustration was too deeply ingrained. I even arranged for them to meet, but it never happened. Even when they were in the same town and knew exactly where the other could be found, they never saw each other.

CLOTHES MAKE THE MAN

I have a friend who used to go dragging for a living. In fact it was a very good living, and he remains a friend even though he's gone back to it.

My friend was uniquely qualified to be a dragger in the modern era. He is quiet, which means he doesn't talk back to the skipper. He is a Brit, so he knows that he's innately superior to the skipper (but doesn't need to talk about it). He doesn't need sleep for weeks on end. He lives on coffee and cigarettes. He's a computer and navigational genius. And he has no moral precepts that prevent him from enjoying the rape and pillage of the sea floor. He is also lazy, which means that he will do almost anything to stay in the wheelhouse doing what he does best, rather than go out on deck where the big waves and steel cables play. His final qualification is that every few weeks his perfectionism and persnicketiness catch up with him, and he spends a day or two blind drunk and happy.

Banks and credit unions don't like people who wear Stanfields and crusted jeans. But my friend didn't have that problem. He was always neat and clean-shaven, showered and trimmed. His problem was that he had too much money. If you go to the bank overdrawn and try to pay your Hydro bill before they cut you off, the bank will find a way to help you out. If, on the other hand, you want to cash a cheque for

$17,000 (three weeks' wages) they don't like it. Even if you have an account at another branch of the same bank, and you want to deposit $12,000 to $14,000, and only take $3,000 to $5,000 in cash, they still don't like it.

"Do you want a cashier's cheque or traveller's cheque?"

"No, I want cash."

"That's an awful lot of cash, sir. Wouldn't you be better off with traveller's cheques?"

"I've never seen anyone use a traveller's cheque in the King's Arms, and I don't plan to be the first."

"Perhaps you'd like to speak to the manager?"

And so it goes. The manager tells you that it would be more responsible of you to put your discretionary capital in a term deposit, and you say "I wanna get pissed." They just don't understand.

A bunch of us were standing around the fish camp one day. Someone needed change. We all laughed. No one had anything less than a $100 bill, and between us we had $30,000 in our jeans. We had all delivered the night before and there hadn't been time to get to a bank. Come back in May, I said. I'll be living out of the jar I throw my loose change in. Money's no big deal. Either you have it or you don't—mostly the latter. But what is a real piss-off is when you have the money and they won't give it to you.

A fisherman walked into a credit union in Town A and asked for some money from his account. His first problem was that his account was not in Town A, but rather in Town B. After some discussion the clerk agreed that this didn't have to be an insurmountable problem, but that it would require a phone call. So phone! Well, we can't actually transfer funds;

you would have to apply for a loan, and then if Town B chose to grant the loan, we could give you the money here, and you could pay it back there. So the clerk phones the branch in Town B and explains the real problem. It seems that the customer is dirty, smells bad and is not the kind of person that the credit union wants to deal with.

Had they asked him, he could have explained that he was dirty because he had been chasing halibut for the last ten days, and he smelled because he had caught them. But they didn't ask him. What they did do was tell the clerk in Town B that she should not even consider giving this person a loan, and they would be very upset if she did so. Since the fisherman in question not only had excellent credit but a large bank account to boot, the clerk in Town B assured the clerk in Town A that the application was a pure formality, and that it would be best if they handed over the cash—post-haste, if not sooner.

Clerk A eventually had to do it by the book, but wrote a long letter to the credit union in Town B, insisting that the terrible customer should never have been given a loan and that the clerk responsible should be declared incompetent. Shortly thereafter, the clerk in Town B took an upgrading seminar on judging loan applicants. The basic rule, they said, was that any applicant who wasn't shaved, trimmed, wearing a suit and polished shoes, should be rejected out of hand. Their attitude was, "if they won't show respect for the institution, don't give them a loan."

Now, I've been a farmer as well as a fisherman, and whereas bank managers in the prairies prefer that you wipe your boots outside, they recognize the smell of cow-shit as the smell of money. Seems to me that fish slime falls into the same category. But I guess bankers here on the coast don't see it that way. If they won't give us money unless we dress and smell like them, then maybe we shouldn't feed them unless they dress and smell like us.

Some years ago fishing went through some very hard times and a lot of boats were seized by the banks. There were a whole pile of them tied up in Port Hardy and the bank wanted them to be in Steveston. There were a lot of broke fishermen all over the coast, many of whom had experience and even papers to run fishboats, and despite some animosity toward the banks they could probably have been hired to deliver the boats south (I certainly would have). The bank, on the other hand, decided that some of their managerial staff might like to drive their new fleet.

One of the managers worked his normal shift and then went down to the harbour, jumped on a very nice yacht-like fishing boat and left for Steveston. About four hours later he sent a mayday. He reported his position as being off Pulteney Point. The nearest boat able to respond was the ferry. It steamed off at high speed and somehow spotted a deck light that was already under water. The crew was saved and the rest of the repossessed fishboats stayed safely tied up in various harbours around the coast. The point is, we don't understand them and they don't understand us. What they do seem to understand is entrepreneurial greed.

Willy the Weasel (not his real name, but it could be) bought a resort tucked away in the back of a pretty little bay. The resort was a sinking, dirty, rotten, ill-functioning mass of slime and rust. It fit Willy to a T. He added some new paint, raised all the prices and put up a big sign on a rock out at the front of the bay to tell everyone he was there. He took out ads in all the big city papers, made some unsubstantiated claims and waited for the money to roll in, which it very shortly began to do. In fact, so much rolled in so fast that he lowered the help's wages, raised the prices again and stopped even pretending to fix the water system.

Still the customers came. Willy couldn't keep help and got no return customers, but he did get one thing right. He had a perfect location. To fix that, he brought in some heavy equipment and proceeded to denude the small valley behind the resort and to blow big holes in the surrounding rock cliffs. The result was to make the place more visible from the front of the bay. He had to put in more floats to accommodate the yachts that wanted to tie up there.

Today the easiest way to find the resort is to look for the tall plume of oily smoke that pollutes the bay and surrounding islands. The previous owner had burned his garbage in the back of the bay in a small clearing made just for that purpose; he had made a small, very hot fire that reduced everything to a bit of grey ash that washed away with the next rain. Willy burns his garbage (flammable or not) on a rock at the entrance to the bay. He pours diesel fuel all over it and sets it ablaze. He's careful to do this at low tide so that the rising water slowly dampens and extinguishes the fire. Smouldering masses of nameless glop drift off with the rising tide and eventually pollute all the tiny beaches in the area.

The other easy way to find Willy is to follow the endless line of boats going in and out of the bay and join the line of people waiting to be used. And yes, of course the banks love him and wish that we would all do the same.

GETTING READY FOR THE OLD DAYS

We opened for sockeye trolling three days ago. This morning we made our delivery and started praying for another chance to earn a bit of money. No matter what your intellect understands, it's impossible to believe that we may only be allowed to fish two-and-a-half days this year. Two-and-a-half days' work for the whole year. We trolled forty-five hours in two-and-a-half days, but that's only because we quit early, hoping to catch up with the dressing and get some sleep. We never quite did either.

When I started trolling some forty-odd years ago, we went out early in the spring and trolled until autumn. We took time off between big runs to go on vacation, attend family reunions and take the kids to Expo. Down in Blackfish Sound we worked the tides and dodged each other in the fog. At night we'd tie up behind the barges in Double Bay, or if we had a big load on, we'd go up to the fish camp in Mitchell bay and deliver.

Whatever came on board we kept—either to sell, eat or give away. One family at the fish camp had so much red snapper given to them that they gave it away to every friend or tourist who happened by. One year I did all my winter canning and freezing from the giveaways of fishermen.

The boats we had then weren't much to look at. Small double-enders, with used and salvaged hardware, and the only electronics aboard were the transistor radio for listening to music in the evenings. If you were serious, you had a depth sounder that moved a needle over a roll of paper and made a tracing of the bottom. They worked pretty well, as long as you remembered to keep feeding them paper.

It was a good life. We even made some money. I fell in love with trolling, and started dreaming about how to get there in my own boat, and how to do it right.

Well, I'm finally ready.

I have a forty-foot troller with aluminum poles, hydraulic gurdies, four radios, two sounders, two radars, GPS plotter and Russell Black Box. My cell phone rings every time my fish buyer wants to share the latest rumour. I'm considering a cell fax and going on the Internet so as to be up-to-the-second on weather, tidal anomalies, fish locations and prices.

If it were only twenty years ago, I'd have it made.

Our prayers and curses didn't do much good—we got another half day, and that was it for the year. Some things just won't compute. I've done it all right. I've spent my money, paid my dues and worked myself to a frazzle. So now I don't even get a chance to try?

Too many things in my life seem to go this way. I'm old enough now to think I understand yesterday. Tomorrow has always been a great blazing hope full of every promise there could ever be. But today—that one's got me stumped. I spent all my yesterdays getting here, but I still don't know how to get from here to tomorrow.

In another era, these trips we make would be epics. Men and women in small boats, leaving a settled home, fighting wind and tide to get to the rocks at the end of the world, to capture and bring back the strange creatures that live there. venomous quillback, brother to the deadly scorpion fish. Or red-fleshed sockeye, which can live in salt or fresh water. We have been there in gales and glassy calm, in fog and on star-clad nights, alone and in company with four hundred other boats. Always pursuing, seeking, hoping. It is a strange game—as old as man.

Scylla and Charybdis are no strangers to us. Like all who seek, we thread the needle of disaster—as much to have been there as to find the goal. Marco and his father, returned as beggars, un-recognized but for one old servant. Or Ulysses, recognized only by his dog. The journey had changed them, almost beyond recognition. And that is the point. Our stories always need the secret hoard of gold or jewels, hidden deep beneath the dirty rags. But they are the metaphor and and— even if real—of less value than the honour and the tale.

There are no winter stars this December night, but wind aplenty and low skies filled with sleet and rain. Up the inlets the freezing winds—sixty knots straight down off 14,000-foot glaciers—hit open seawater and glaze rocks, logs and boats with a deadly layer of gleaming ice. Turn Glendale corner—and die.

But here on our sand and gravel island, twenty miles closer to the open sea, it is rain and sleet and salt water warm enough to wash free the bits of ice that form in the corners. The harbour has its thin layer of fresh water ice, broken and gone as fast as sun or surge can touch it. This island, a temporary gravel bar left over from the last great ice sheets, eroding at the edge of the sea. A few miles farther west, and it would never have been. A few miles farther east, and it would be as permanent as the hard stone from which it was caved.

When I cut wood for lumber or carve shapes, I often look at the pile of chips, so numerous and varied, and wonder why I made each one. And then I compare them to my "work" and wonder which is more meaningful. My island is a heap of chips carved from the hard stone islands and swept here by the glacier. I burn my shavings and chips to keep warm on nights like this.

The radio says it will be much the same tomorrow and the next day. More ice and wind to eat at our gravel shores; it will be at least another week before our winter stars appear again. Pleiades, Hyades, Taurus and Orion, steel bright in black velvet, with innumerable bright grains of sky sand around them. Who has scattered these chips in the tides of creation, and who will sweep them up to burn in winter's fire? Cosmos and cosmogony in grains of sand, wood chips and stars.

EPILOGUE

I'm an old guy. In 1976 I was on a trip to Port Hardy to visit a teacher friend when my wife and I passed a road sign indicating the distance to Sointula. I had heard about Sointula all my life. My mother's family had tried to move there in 1917 to join the utopian colony. That didn't work out for them but when we were halfway across on the ferry we decided to move there. I've been here ever since.

Life in Sointula suited us just fine. The community and culture were familiar and I could speak enough of the Finnish language to get along. Soon I was working on fishboats, mechanicing, odd jobbing, shake blocking and doing whatever needed doing to raise a family.

My wife and I bought some land, cut down some trees and built a house. I set up a wood shop and started building small boats and furniture. I also took up wood carving and did fairly well as an artist in my spare time. The community was very rich in those days and local people supported the local arts very well.

I have always loved boats. I had spent my whole life on sailboats, freighters and small skiffs, but in Sointula fishboats were the way of life. I started crewing on local boats. I tried trolling and gillnetting and seining and spent some time crabbing. It was clear from the beginning that I wanted to become a troller. The seventies were the glory days of fishing and while I could get jobs deck handing, there was no way I could ever afford a boat or licence for salmon. There was however a licence available for cod fishing. It cost $50. I got one and fixed up an old dory and went looking for fish. A friend introduced me to a fish camp in Blackfish Sound and as things turned out I stayed there for more than twenty years.

I built myself a new dory powered by an 18 hp Chinese diesel and named it Jigger. Our fishery was a handline jig operation. Being in a small dory (29 feet) means getting shaken a lot and it was always dancing around in the chop. The name stuck and for many years my radio name was "Jigger Jon."

One day while I was out fishing I got a radio call telling me that an old friend was in the area and looking for me. My marriage had dissolved a few years before, the kids were in University and my high-school girlfriend had got in touch to say that we had some unfinished business to attend to. We did so.

In 1995 Linda arrived with all the pent-up energy left unspent for twenty-five years and more money than I had ever seen in my life. We got married and bought a salmon boat. The glory days of salmon were well past but combined with the cod fishery there was a reasonable living to be made. We got a great boat, *Open Sea II*, and fitted her out in fine style. We had great electronics and the best equipment for trolling and cod jigging. I was as happy as a person could hope to be. Linda was also an exceptional artist. We spent our non-fishing time creating art and planning tours of the big seasonal craft shows.

We were sitting around at home one day when Linda looked at me and said, "I don't want to be married to anyone." The next year was a very bad time for me. I was an emotional wreck and didn't know where to turn. Eventually things worked out and I was left with my home and boat and no money.

I had the tools to start over but not the will. I was essentially broken. Between the breakup and the PTSD from my volunteer medical work in Vietnam I was empty. I kept fishing but it was hard to care. The Department of Fisheries kept changing the rules and telling lies. I was afraid that I would damage them or myself if I kept dealing with them. I sold out and at sixty-three retired in very poor mental condition.

I had always wanted to get involved with music. My partners and kids always discouraged me from doing so. I bought a guitar and started teaching myself to play. My musical friends kept telling me I was doing it wrong. I kept doing it. I started doing vocals with local players and performing solo. Eventually I was accepted as a musician and continue to this day busking and performing with friends and strangers that I meet.

Throughout this journey I have been writing. Mostly poetry and short stories about things that somehow seemed significant to me. As a kid in school I found it easy to get extra credit by writing stories about the things we were studying. As a young man travelling I would write forty-page letters and keep a journal. All through middle age I recorded the stories that presented themselves to me or described the images that came into my mind. It hasn't stopped happening. I have to force myself not to write.

I am still here, living in a shack on the beach. I like my home and I love company, especially young folk. There is a great deal of art and music that happens here. When the weather permits I go busking in town. The local people who know me seldom throw any money in my hat, but tourists (in season) provide some change and occasional meals. I will be happy to play you a song or read your palms or cards. I don't charge up-front, but if you want to pay me afterwards I will take whatever you offer and thank you sincerely. I have been paid as little as 35c—not counting people who paid nothing—and as much as $200. There was even a time when I gave a woman $5—because she said she was broke and needed a beer.

When people ask who and what I am, I tell them that "I sit on lonely beaches and listen to whales." That seems to be enough explanation for those who pass by.

TYPES OF FISHING VESSELS AND THEIR CREW

Trollers start out as normal enough boats, prettier than most, but really just like all the others. Thirty-five to forty-five feet long, nine to fourteen feet wide, and four to six feet deep. Then they put their gear out. Suddenly they are 130 feet long, 50 feet wide and 250 feet deep, and they cannot turn sharp, back up or stop. It is at this point in their life cycle that trollers congregate in highly localized groups, and get as close to one another as they dare. There are good and simple rules for avoiding collisions and entanglements, but unfortunately ten percent of the trollers don't know them and another twenty percent are using different rules. Things do get interesting at times.

A sample of troller speech:

"I'll give her a little tickle and you can hug my pigs."

Or to put it in the vernacular, "I'll speed up a little and you can slip in close behind me." Now if the fellow slipping in were to actually kiss your pigs, well that would of course cause problems. Touching isn't allowed.

Gillnetters don't like to be close to each other. They always claim a quarter mile of free space around their net and they always claim they're not getting it. They further claim righteous retribution for any perceived infraction, which they swear they will collect somewhere, some day.

A gillnet is a delicate device intended to be invisible to fish, so that they will blunder into it, entangle themselves and drown. In fact this sometimes happens. But it isn't always the desired fish that gets caught in the web and drowned. And in addition to fish, logs and seaweed love to entangle themselves, much to the detriment of the net—but that's life.

A sample of dialogue exchanged between gillnetters:

"Back off, asshole, you're corking me. You're right over the quarter-mile ring."

"I can't help it; the tide's pushing me to the beach and I'm towing as hard as I can."

"Well, pick it up then." That's gillnetter talk for "I don't like you."

Interestingly enough the people who run trollers and gill-netters have social habits exactly opposite to the way they fish. Walk into a pub near a fishing harbour and look around. Somewhere near the middle of the room there will be a thick knot of serious drinkers. They will be sitting shoulder to shoulder, all leaning toward the centre, listening to someone telling a joke or story. These are the gillnetters. Spread out around the perimeter of the room, as widely spaced as possible, will be a handful of solitary drinkers. If there are two of them together they will be sitting with several chairs between them, and only one of them will be talking. These are the trollers.

In anchorages up and down the coast a small bay that will only hold three trollers will easily accommodate fifteen gillnetters. I've seen die-hard trollers run back out of a pro-tected anchorage to anchor in a gale, rather than face the terror of spending the night within one hundred yards of another boat. This same boat will think nothing of crowding you onto the beach, or holding a cross tack through a big fleet in the fog. A troller doesn't really care if he rams you or not, but he certainly won't anchor within sight or sit next to you in the bar.

Gillnetters on the other hand are sheep until they get their net in the water. Then they all consider themselves to be the last of the rugged individualists, and feel honour-bound to defend their territory and reputation.

Seine boat skippers fall into various categories, but all of them that I have known well have intoned the same mantra. If you listen to the undertones in their constant mutterings you will hear them chanting "Sex, speed, power, money, sex, speed, power, money." Some of them alter the order of these prime concepts and the odd one will add "abuse" as a primary concern, but basically it's all the same.

In anchorages, the seine boats usually tie up to whatever boat happened to get there first. The original boat is probably broken-down and won't have had the chance to set its anchor, so in the middle of the night all ten boats tied together will sweep through the anchorage, entangling everything in their path. Each skipper will blame all the others and be willing to fight anyone who says they're wrong.

In the pub, seiners just naturally assume that they own the place and will be pleased to tell you that they could easily buy two or three such bars out of their crew share. In fact, maybe they did buy it last year and they've just forgotten. A wise man once said, "Where seiners congregate, peace and harmony go for a shit." He was a very wise man, especially in that he never said such things where a seine skipper could hear him.

For some serious fun and deep insight into west coast culture, you should try to arrive at the Dalewood or the Seagate on an afternoon when fishing has just closed and a bunch of newly laid-off loggers arrive from camp. The intercultural dynamics will take your breath away—and probably knock your beer over at the same time. I have it on good authority that any bar in Rupert will provide the same show on a slow afternoon, but I've never had the nerve to go see for myself. I have seen the scars, crushed hands, broken teeth and brain damage—but that just goes with the territory. Attendance is voluntary, not mandatory.

A GLOSSARY OF SOME USEFUL TERMS

beach knot: The knot used to tie the end of a seine net to the beach. It must be possible to untie this knot in a controlled manner, despite the fact that the line is stretched to the breaking point. The popular manner of learning this knot is to tie a heavy line to the back of a 4×4 and then run to a tree and tie it up as the 4×4 drives away. Then you have to untie it while it keeps pulling. FUN! If the knot jams, or won't come undone, you have to cut the line.

beach man: The person who gets to run up the beach pulling a vast length of heavy line behind him. His job is to tie it to a secure spot and untie it later when the skipper signals him to do so. The beach line gets a lot of stress and the beach man has to be very careful. Other than that, he has a lot of free time.

braille or **brailler**: Either a small net used to transfer fish from one place to another, a small dip net to transfer a handful of herring, or a six hundred pound basket used to offload salmon. When a net in the water has too many fish to be lifted onto the boat a brailler is used to pick up a few at a time.

cash buyer: A packer who works on his own and offers direct cash payment on the spot. He makes his money reselling the fish to various factories. Some cash buyers resort to herbal or chemical inducements to persuade fishermen to part with their fish for less than the going rate. *See also* packer.

drum: A very big aluminum spool used to roll up nets or long lines. The drum is sized to suit its intended purpose. They are very powerful and will wind up anything that gets caught in the net or line—crew members not excluded.

gillnet: A net designed to be invisible to fish. Contrary to popular belief the fish do not get caught in the net, but rather in the holes in the net. If the holes are too small the fish just bounce off the net. If the holes are too big the fish swim straight through. So the fisherman has to have a number of nets, and then guess what size mesh will get the most fish. Gillnets collect every bit of trash and weed for miles around and are occasionally destroyed by schools of dog fish.

humpies: the smallest and most common variety of salmon, so-called for the humps the males form once they enter fresh water to spawn. More commonly known as "pinks."

lazaret: The area in a boat referred to in old novels as the "steerage." It is the area around the rudder shaft and while it is probably the part of the boat that has the least motion, it is usually low, dark, damp and noisy. In fishboats it is usually used to store gear that is absolutely essential to have on board, but that you hope you never need to get to quickly. The name comes from the most famous passenger ever to use such quarters, Lazarus.

packer: A boat that hauls fish caught by other people. The packer usually works for a fish company. They will meet you on the fishing grounds and take your fish in so that you can continue fishing. They pay less than the rate at the factory but they also provide ice and a few supplies. *See also* cash buyer.

running line: A rope rigged in such a way that the far end of a seine net can be pulled back to the boat by a deck winch. This line floats on the surface and is a favourite target of sports fishermen who, having failed to notice the net in front of them, run over and sever the running line shortly before hitting the net and destroying their propeller.

sea anchor: A canvas cone about three feet across at the large end, held open by a steel hoop. It is used to stop the end of the net so that the boat can drive out from under it. It is designed to have the maximum resistance to being towed through the water, yet still be light enough to be easily handled by one man. It's the skiff man's job to throw the sea anchor when the skipper gives the signal. One skiff man of my acquaintance got his foot caught in the sea anchor line and launched himself—on two separate occasions. If you're stuck in the skiff and it's cold and wet you can pull it over your head and stay a little dryer.

scotchman: A pneumatic balloon made of heavy rubber, between twelve and thirty-six inches in diameter. Universally used as a float at the end of nets or other fishing gear. Also used as bumpers between boats. (I have always assumed that the term scotchman refers to their bright red colour, like a tam-o'-shanter hat.) There is the apocryphal story of the fisherman who spotted a few scotchmen drifting free. Upon investigating, he found a drowned woman tied to the floats. He called the RCMP and reported that he had found a drowned woman tied to a scotchman. After a significant pause, the RCMP responder replied that he understood how you could tell that one was a woman, but how did he know that the other was a Scot?

seine net: A seine net is a great huge heavy thing that requires a sixty foot boat to carry it and a vast array of heavy machinery to operate the thing. It drops into the water in the form of a letter C, then closes on the top (to become the letter O) as well as closing on the bottom. As the net is rolled back onto the drum, the fish (if there are any) are forced into a bag at one end and then "rolled" onto the deck. If there are too many fish for the machinery to roll them onto the deck the bag can be assisted aboard with a hydraulic single fall hoist, or the bag can be emptied with a brailler.

set: The act of putting the net or line in the water. Also the act of having the gear in the water, as well as the result of having the gear in the water. "We set, we watched the set and we had a good set."

beach set: Tying one end of a net to the beach, usually to a tree or rock (although I do know of one small lighthouse that was pulled off its base and lost into deep water due to a seine boat using it as a tie-up point). The tide pushing on the net plus the boat powering against the tide creates massive forces that often enough result in broken gear. Such breakage can and does result in serious and sometimes fatal injuries to the beach man.

open set: Putting out a seine net so that it drifts in the tide. Stress is greatly reduced, though fish catch can also be smaller. In the old days the skiff man and the beach man sat in the skiff during open sets then rowed the beach line back to the boat. Nowadays the running line is used to pull the net back and there is no need for the skiff to be used on open sets.

skiff man: This is the lucky fellow who rows the beach man to the beach and later returns him to the boat. He has various other duties, including scaring fish back into the net and keeping debris out of the net. His most important job is remembering to put the drain plug back in the skiff so it doesn't sink when it's launched.

troll: A mythical creature similar to a seine skipper. Also a system of catching salmon wherein you pull sixty to one-hundred-and-eighty lures behind a boat, while attempting not to get them tangled.

ABOUT THE AUTHOR

Jon Taylor is a retired fisherman and boat builder who has lived near Sointula, on Malcolm Island, for many decades. He has been a lifelong writer of poetry, memoir, essays and fiction, and is also an avid musician.